An Honor and A Guard

From Omaha Beach to the Heart of Nazi
Germany–One Infantryman's Journey

Praise for *An Honor and A Guard*

"Riveting and thrilling true account of a young man's treacherous path from the beaches of D-Day into the depths of Germany. It's in the vein of the best of Band of Brothers or Ken Burns. Highly recommended tribute to the greatest generation! Must Read!"

— Gregory W. Siewny MD
Chairman of the Board of Trustees
Atrium Medical Center

***

"Dr. John Birch offers an excellent overview of an individual soldier's journey from induction in the U.S. Army to Germany during World War II. That soldier was Dr. Birch's father, Sergeant Forest S. Birch. This is as good a description of what the experiences of young Americans, 18, 19, or 20 years old, were in World War II. Sgt. Birch's entrance into combat began on D-Day at Omaha Beach. He became a soldier two years earlier.

The personal aspect of this work is what makes it so exceptional. Dr. Birch mentions, especially in his initial pages, that his father was reluctant to discuss his wartime service. Veterans rarely do. They do not want their children to experience, even second hand, the horrors they saw in war. It would not be until June 6, 1994, the 50th anniversary of D-Day that Sgt. Birch truly opened up to his son and did a long video interview. That interview is printed in its entirety at the book's conclusion.

The transcript of Dr. Birch's interview with his father is priceless. This book is priceless... a MUST for anyone interested in history or the triumph of the American spirit."

— Dr. Robert Young
Professor
American Military University

***

"In this highly readable book, Birch draws the reader in on a journey to a front row seat of raw emotions, unheralded heroism, and chance encounters with generals like George S. Patton."

— Jeffrey C. Stone, Ph.D
Chair, Department of Educational Leadership
Associate Professor of History and Education
American Public University

An Honor
and A Guard

From Omaha Beach to the Heart of Nazi
Germany–One Infantryman's Journey

John F. Birch, Ph.D

Illustrations © 2024 Brynne Davis

ISBN: 978-1-958407-25-7 (Hardback)
ISBN: 978-1-958407-26-4 (Soft Cover)

Book design by designpanache

ELM GROVE PUBLISHING

San Antonio, Texas, USA
www.elmgrovepublishing.com

Elm Grove Publishing is a legally registered trade name of Panache Communication Arts, Inc.

Table of Contents

This work of love and devotion is dedicated to my parents, Forest S. Birch and Betty E. Birch, part of "the greatest generation" who endured hardships from which my generation was largely protected.

It is also dedicated to the family my parents established in their life together since their marriage after Dad returned to the States following World War II.

In addition, this work is dedicated to the men and women who served throughout the years in the armed services in defense of the United States of America. The sacrifices these people make are not measurable yet are displayed through the quality of life U.S. citizens enjoy.

Preface

YEARS AGO ON my dad's birthday, July 13, I decided to take on something I longed to do but was also somewhat hesitant to do for various reasons, which is to tell my dad's story of his experiences during World War II, from his training in England, landing at Omaha Beach, on through France, and then into Germany. The title of this work comes from the Latin inscription on the insignia for the unit in which Dad served, which was the 175th Regiment of the 29th Infantry Division. The 175th and the 29th each have an insignia. The 175th insignia includes the Latin words "DECUS ET PRÆ-SIDIUM" which translated means, "An Honor and A Guard." However, it was not for the insignia alone that I chose to use this for the title, but rather because these two qualities are indicative of who Dad was as a person and more personally what he was for my mother, sister, and me. Growing up in his household, I watched Dad honor his word and commitment to others and guard his family in his character, moral standards, and work ethic. Personally, I always knew Dad to be a man of his word, and I could always depend on what he told me.

Even when he had not made a previous commitment to others, I saw Dad honor his own character by sacrificing his time and comfort to go help someone who had a need. There are many instances when I saw this in action but will mention a couple that were close to home. One winter when my first cousin, who lived next door, was out in the snow under his old car trying to get it to run, Dad was home relaxing in our house, but he watched through the glass sliding doors in our family room as my cousin worked and worked. In a few minutes Dad got up, bundled himself against the cold weather and went outside to help him. He was not asked to do so but he saw the need and it called something forth from him to help his nephew when it seemed he was not making much progress.

Dad also took the initiative when his namesake, his oldest nephew on his side of the family, had essentially become estranged over time from his own parents. Though he was named after Dad, we all knew him as Buddy. Over time Buddy felt he had to sever ties with his mother and father to make his own way in life and avoid the abusiveness with which he had been raised as a youngster. After he was married and had his own family, Buddy would drive back and forth to work, passing by his parents' home day after day without ever stopping to visit. Buddy not only shared Dad's first name but also Dad's commitment to work hard to make a good life for himself and his family. When Buddy's mother, my dad's half-sister, became very ill the doctors at one point did not give her long to live. Dad spent time with her and helped her come to terms with what she was facing, but most importantly about her relationship with Jesus Christ. One day Buddy received a call from my dad telling him that he needed to reconcile with his mother and forgive her any wrongdoing whereby he had been hurt. He also told Buddy that should he leave things as they were and his mother pass away, he would never forgive himself. Buddy took Dad's words to heart and was able to become reconciled with his mother. This meant everything to his mother, and Buddy was grateful to Dad ever after. My aunt was able to die in peace. Some years later when Dad was facing open heart bypass surgery Buddy, who had undergone the same type of surgery earlier on, called and came by to sit and talk with Dad about what to expect. What Buddy shared was extremely helpful for Dad, and for the rest of us. In many ways Dad was more like a father to Buddy than an uncle. I could go on and on about relatives who Dad helped, with whom he talked and prayed, and upon whom he left an indelible impression. His purpose in all these encounters was not to impress but genuinely to help.

Without any doubt Dad was my guard. On a couple of occasions years apart when I was stranded and unable to get my own car started miles from home, it was my great relief to see my dad arrive, get out of his car, and help me get going again. Before he even touched my car, just having Dad there made me realize that I was no longer stranded. Even beyond the more common things like rescuing me when my car broke down, he rescued me by teaching me the way that is right. I say more about this in the final chapter, "Reflections and Musings." Both Mom and Dad became Christians when I was still an infant. While my sister remembers the change that occurred at home, I never knew anything but rearing by Christian parents who taught me, by word and example, the way to live life. In addition, when the time finally came, they both also taught me how to die. Their lives modeled what it means to be the person that God intends you to be. Dad was my guard to keep me from going down paths in my life that would lead to ruin. He and mom were not afraid to discipline, because they knew that you discipline those you love, not those who you care nothing about. Without any doubt I can honestly say that every time I was disciplined by them, I felt loved by them. What is interesting about my relationship with my dad is that we both tended to be quiet, not engaging in constant small talk like some people do. It is not that one way is better than the other, it is just that Dad and I were both turned that way. There were times when I was going somewhere with Dad and hardly any conversation was exchanged between us. Still, that did not matter because I felt loved just being with him. This has taught me that presence with the ones you love is all important. Even if you do not have words of wisdom, or any words at all, just being with your loved ones is what counts. Time in their presence matters, time in your presence matters to them.

Earlier I said that I longed to tell the story of Dad's experiences during World War II, but I was also a bit reluctant to make the attempt. Dad left me two gifts to help me do this. First were all the times over the years when he gave in to my requests for him to talk about his part in the war. I was probably seven years old when I started asking, maybe even begging, him to tell me about it. He would never talk about it, except on occasion when I kept asking and then only when he, mom, my sister, and I were at home alone together. Thinking back, I feel bad about not realizing that talking about it was most definitely not easy for Dad. As a kid thinking about myself, I never considered the emotions that remembering would stir up. It did not dawn on me that remembering could be very difficult by forcing Dad to relive experiences by way of the emotions he felt. Yet he kept giving me those stories to the point that I could remember them by rote.

The second gift came on the 50th anniversary of D-Day. On June 6, 1994, Dad consented to make a video at home recounting his experiences during World War II. This time he told his story to my mother Betty, his granddaughters, Kindra, Katrina, and Kara, son-in-law Roger and my sister Jeannine, and my wife Debbie and me. His family had grown. This time he had a larger group wanting to get a more permanent record of his story. As he sat in his chair and began to talk, all eight of us sat as close as we could around him, some on the floor, others on the hearth, barstools, or kitchen chairs, totally mesmerized by his story. He answered the family's questions, never letting on that each memory carried with it associated emotions, and many of those memories/emotions were not good. In these ways Dad gave me the tools to tell his story, or it might never have been told. While it took some time for me to be able to do it, later on I sat down with the video Dad made and transcribed it to have a written record in his own words, the way he told it. I found that I wanted to read him as well as listen to him. My daughter Kara used Dad's story to write a paper in high school reflecting on World War II and how her grandfather personally experienced the war. She won a scholarship for her work on this paper and the significance of this story.

After Dad passed in 2002, I tried for the third or fourth time to obtain his military records, which were not available from the National Archives. I always kept running into barriers and it was frustrating to read accounts of other soldiers and their units, only to find out that about twenty years of military records held in the National Archives were lost in a large fire, preventing me from obtaining Dad's records. Still, the archivists had Dad's medals record on file and were kind enough to reissue all his medals since they could not provide the records I was requesting. Three of these medals are personalized with Dad's name inscribed on the back. The three personalized medals are the Purple Heart, the Bronze Star, and the Good Conduct Medal. I did not know these medals were personalized, although I probably should have since they are individual decorations. When I opened the boxes that contained all of Dad's medals tears came to my eyes. I wished so very much that Dad had been with me for that. He would have been especially appreciative of this because most of his original medals were lost in a fire at my grandfather's house many years earlier, even before I was born. I would love to have seen the look on Dad's face when opening the new medals, reissued and with all their associated ribbons.

Subsequently I hired a group of researchers, former military personnel, to obtain the military records from Dad's unit, which included the daily morning reports of events. This allowed me to assign specific dates and locations to the stories Dad gifted to me over the years. This was a great help even though there were still a couple of events which continued to give me trouble as I worked to pinpoint them on a calendar and various maps. These morning reports proved invaluable for enabling me to develop an accurate chronological accounting of Dad's tour of duty. The difficulty with these records was that they were kept on the fly, in the heat of battle on the front lines where the fighting was going on daily. Some of the reports are missing specific details and lack a legible quality in the typing and writing. Since I had the stories in my head and knew the details, I used the dates on the daily reports from Dad's unit to align his accounts with maps in books which follow the trek of the 175th Regiment. When I started writing, my dream was to be able to have his account published. At the time it did not dawn on me how emotional it would be to write his story. In some ways I still feel like the seven-year-old sitting at his feet listening to him relating his experiences to me.

1. A Metronome at Sea

"I CANNOT BELIEVE the entire cycle took twenty minutes; it actually took a full twenty minutes!" he thought to himself. The ship was large, no, more than just large, it was enormous. It seemed very much like a floating city.

While the RMS *Queen Mary* was originally built as part of the Cunard line of ocean liners, her mission now was to serve as a workhorse troop ship for the Allied forces, making treks back and forth across the North Atlantic Ocean between the United States and the United Kingdom. Earlier, this giant shuttle of the seas discarded her original colors and donned a light gray designed to make her nearly invisible on the water, less of a target to German submarines, called U-boats. This new coat gave rise to a nickname for the ship – the "Grey Ghost." Additionally, a major refitting effort enabled the Grey Ghost to accommodate 15,000 troops.[1]

However, it was only possible to carry this number of soldiers across the ocean during the summer months when the waters were calmer. Otherwise, rough, choppy waves could result in troops being tossed about, resulting in broken limbs and other injuries. After her makeover, she began making a significant contribution to the build-up of Allied forces prior to their invasion into Europe. Essentially the ship could carry an entire division across the ocean in less than six days. Her speed was a welcome characteristic in the event she would have to make haste to outrun one of the U-boats. Now that she was permanently assigned to the European theater, the giant vessel became a magnificent "GI shuttle" for ferrying thousands of American and Canadian troops across the Atlantic.

On this day in June of 1943, she was destined for Gourock, Scotland, which was located down river a little over 32 kilometers (20 miles) from where she was originally built and launched back in September of 1934. Just a few days past, she set out from New York packed full of U.S. troops going to England for focused training prior to their entry into the European theater, though the troops were not yet aware of the details of their mission. Among the thousands of soldiers making this trip was Private Forest S. Birch. For this trip he had volunteered for Kitchen Patrol (K.P.) duty so he could get a personal bunk, not because he was proficient in the kitchen. The truth is he had never really been interested in kitchen duties of any kind. Soldiers traveling on this behemoth of the waters spread out and sought every available space to be found, like explorers on a hunt for hidden treasure. For a trip such as this, space and personal space in particular, became a coveted treasure. There was no room to lodge the enormous crowd of troops without making special provisions. Even then there were limitations. For this situation the special provisions involved the use of "standee bunks," which were used in place of the ship's hammocks installed earlier on. Standee bunks were essentially tiered sleeping spaces stacked up to six feet high, like empty storage shelves waiting to be stocked. These shelves were going to be stocked with soldiers who would pile themselves on, along with all their personal equipment, for the purpose of getting some semblance of sleep. If it went well, one might even feel rested after waking. With only an eighteen-inch clearance between each bunk, this arrangement did nothing at all for those with claustrophobic inclinations. It was also problematic for those sleeping on the bottom when those higher up on the stack happened to yield to the throes of seasickness. It was a difficult situation for getting rest.

Moreover, soldiers using the standees slept on a rotation basis, sharing their designated spot with other soldiers. Even the standees did not allow each person to have his own personal sleeping space for the entire trip. The rotation system allowed each one to sleep inside for two nights and then outside for two nights. This meant that after sleeping on a standee for two nights a soldier had to pack up all his personal belongings and equipment carried on his person and move to the top deck of the ship. At any given time about one-third of the troops would be sleeping outside on the open top deck. Sleeping on deck was only feasible in the summer months, as winters on the North Atlantic were much too cold. Many soldiers preferred sleeping outside because it was not as dank and claustrophobic as it was below deck. Perhaps more important than this was the reasoning, correct for the most part, that those who were on the top deck, or "topside," had the best chance to survive should the ship be attacked and torpedoed by one of the German U-boats that continually prowled the North Atlantic like great roving mechanical sea creatures searching for vulnerable prey. In addition to the enemy from below, the weather from above often quickly became an enemy causing great difficulties for those who had to take their turn topside.

Word came down from command headquarters that a certain number of men was needed to serve on K.P. duty, and that those who volunteered to serve would be compensated. As it turned out, volunteering to serve on K.P. duty meant Birch would be compensated with his own bunk that he did not have to share for the entire trip. This was treasured compensation indeed, and one worth performing kitchen work for the length of the trip. Having your own bunk was of high value and therefore, during each trip across the Atlantic Ocean, the U.S. Army always managed to get the needed personnel to perform the crucial kitchen services for their brothers in arms. Birch's duty was straightforward. He collected meal tickets from his fellow soldiers, went to get their meals, and brought the trays back to the tables where the soldiers were seated. After everyone finished eating, he then collected the empty trays, wiped off the table for the next group, and repeated the process, over and over until everyone was finished with that daily meal. That was it. Each day at mealtimes he was busy delivering food and collecting tickets, delivering more food, and collecting more tickets…on an on it went until everyone was fed. He did this for two scheduled meals a day offered using six staggered sittings – breakfast was served from 0630 to 1100 and dinner was served from 1500 to 1930. During the meals soldiers were allowed to take as much food as they wanted but they had a limit of forty-five minutes to get through line and finish eating their meals. As one group left, another group was already lined up waiting to be served. As they exited the dining area, the troops were permitted to take sandwiches so they might have something to snack on between meals. Apart from performing this duty Birch could move about as he wanted in his designated section, but there really was not much room to maneuver. The main deck was crowded with others like himself on their way to fight in the Second World War.

To help manage the overcrowding as best as possible the ship was divided into three vertical sections, designated as "Red," "White," and "Blue." The Red section ran from the ship's bow back to the number three stairway but did not include the sun deck. The White section covered everything between the number three stairway and number four stairway, including the sun deck. Then the Blue section ran from the number four stairway back to the stern of the ship.[2] Every soldier was given a button, colored to correspond to his assigned section, and it was forbidden to visit other sections. When he was not moving about his part of the ship, which did not occur often, Birch could at least try to get some rest in his own personal bunk. He was glad he had volunteered to serve food. It was a major comfort to have his own spot where he could go, rest, and be left alone to think.

Currently he was feeling the ravages of seasickness. As he lay there, he decided to time the ship's slow rocking motion as it sliced through the cold waters of the North Atlantic. The side-to-side movement was continual, a regular back and forth motion like a slow, steady metronome keeping time as it measured the cadence of his journey. Only this metronome of the seas continually measured the time and distance from home. Each completed cycle meant that Birch was eleven miles further from home and twenty minutes closer to his destination. He timed this cycle from vertical – over to port – back to vertical – over to starboard – back to vertical. The complete cycle took twenty minutes. Why he made the effort to time this he did not know. It seemed like something to do at the time, perhaps in the hope of taking his mind off being seasick. He would never forget how long the cycle took, and how slow and perpetual the motion felt. This slow procession back and forth managed to weave its effects into the depths of his being, as it had for many others, to bring about the misery and discomfort that he was feeling at this moment. It seemed like it would never end.

This experience was not uncommon and added to it were things that only worsened the misery. For one, the soldiers on board the ship only washed infrequently. This was not due to uncleanliness, but rather in large part to the lurking fear that one might be caught in the shower if the ship was attacked and torpedoed. The infrequent bathing was also due in part to the limited availability of facilities on a ship designed to carry at most 2139 passengers and 1101 crew. Today she carried nearly five times that capacity. The common wisdom on board was that it would be better to be dirty and safe than to be clean and injured or killed. This resulted in the production of an atmosphere consisting of a strange mixture of odors including cigarette smoke, sweat, vomit, and diesel oil. It is very possible that if the motion did not make one seasick, the bizarre cocktail of odors would. This was the nature of life on board the Grey Ghost.

As he lay there, he thought of home, of his dad Levi, and other members of his immediate family, which included his stepmother, half-brother, and half-sister. Levi J. Birch was a stocky man who wore wire rimmed glasses and smoked a pipe. Some would have said he bore just a slight resemblance to Harry S. Truman. His second wife, Garnett, was slightly shorter than her husband and heavyset in build. Forest never really had the opportunity to get to know his biological mother, Mae Birch, who died when he was just a child at the tender age of two. According to her picture, Mae was a very thin, frail woman, and when others looked at her photo, they could see something of her in her only child's face. He had a picture of her and over the years could not help but try to pull up some memory or image of her. Forest's aunt Maude, who was Mae's sister, gave him a couple of keepsakes that had belonged to his mother, and over the years Maude was a source of information about Mae. What little he came to know of her he attributed to his Aunt Maude.

His current memories would not be complete without the mention of Tippy. Prior to leaving the States, Forest posed for a picture in the front yard of his family home, decked out in his dress Army uniform stooping down in the yard with Tippy, who hung close at his side. By all appearances Tippy looked like any family dog might look out on a farm. The Birches were not farmers but rather well drillers, and in particular water well drillers. They lived in the country in a very modest dwelling and Tippy became an integral part of the family, enjoying many things that people did. However, this normal-looking family dog had an unusual quirk. Probably the greatest enjoyment for him was riding in the car. He liked it to a fault, so much that he was routinely banished from the car. Still, he became very astute at knowing when someone in the family was getting ready to leave and wanted to be sure to get his seat. To this end Tippy resorted to a solution, which was to quietly slip outside and conceal himself on the back floor of the car. It did not matter who was leaving, Tippy just wanted to take part in this unique and fun way of getting around. He wanted to ride.

When it worked as planned, Tippy would lie in the back floor of the car and make no sound whatsoever. He would wait, bide his time, and give the driver, usually Levi, plenty of time to get far enough away from home such that it would be too much trouble and take too much time to go back home and restart the trip. Then at exactly the right moment, with a flair all his own he would jump up, place his paws on the back of the front seat, tail wagging, and enjoy the ride. Of course, this was a total surprise to the unsuspecting driver, again usually Levi. The magic of it was not so much that he stowed away, but rather his sense of timing. Forest was intrigued by Tippy's behavior and lingering in his mind was the question of how the dog was able to time things so well as to prevent a return to the house and banishment from the automobile. This was a mystery of this canine that he had never solved.

Forest thought of his younger half-brother Millard Birch, known to the family as Mac. They were close and ran around the town together back home. Millard was a bit more of a rough neck than Forest, and more than once he helped get them out of fights, as well as into an occasional fight. The two Birch brothers were close and complemented each other well. They had experienced a lot together in Forest's twenty years of life. Mac knew he would soon be following his older brother to serve in the armed forces. Forest held onto the hope of being able to survive the war while serving in his own role in the infantry as a member of a 57 mm anti-tank gun crew. Then he could return home where he and Mac had worked together, double-dated together, and spent many a weekend taking in the town together. Mac was somewhere else back home, likely absorbed in his own thoughts of the war, his future role in it, and his older brother. Currently in his own mind Forest wondered how much he might be changed if, and when he made it through this war, did his part, and was able to go back home. Questions came to mind such as, "Will I be much different, or pretty much the same person?" and, "Will I be whole and able to work and get around on my own?" Hidden beneath these more obvious questions was a lingering concern that the war could change him at a deeper level, more to do with his character. His hope was that one day he would marry and raise his own family. He wondered what kind of husband and father he would be. He continued to think of his own dad with whom he had a good relationship. Though he was the only child of Levi and Mae and was never able to know the experience of being raised by his biological mother, he considered himself fortunate in that he was raised in a family with a father, mother, and two siblings. That was not always something for which he was grateful, especially during his teenage years back home in the country.

He recalled his days as a young teenager when he often heard the familiar whine of his dad's truck engine from some distance as he headed home. The weekend was the part of the week he most dreaded because that was when he expected his dad to be drunk. On hearing the truck's engine nearing home, Forest ran out into the corn field, hid among the stalks, waited for his dad to arrive home, go into the house, settle down a bit, and then fall asleep. This was an uncomfortable thing to have to do on cold or chilly evenings, or when it was raining, but it was a matter of waiting out the effects of the liquor and the chance of being the target of Levi's drunken anger. Sitting among the plants, listening to the soft rustling sound as the evening breeze played with the leaves on the stalks, gave him the opportunity to think and dream. As a boy who was quickly becoming a man, he had goals and aspirations, but kept these close to himself. It was a difficult thing to be too trusting of others, especially those who had let him down. The cornfield for him became "life's workshop" where he imagined his fantasies, fretted about the turbulence of his youth, dreamed of his future, and came to terms with the realities of life. Eventually though, he had to tiptoe back into the house in the hope that he would not disturb Levi. He hated to see his dad like this, and even more, he hated what drink did to his dad. For Levi tended to be a mean drunk and likely to strike out in ways that were not characteristic of him as a man and father. That was now long past, and Levi had overcome this period of drinking in his life. He became more compassionate, and more tolerant of his children's mistakes and mishaps, which sometimes required him to yield and consent to do things for his kids that required a great deal of patience and understanding. An example of this came to mind as Forest lay there in his bunk waiting for the next round of mealtime duties.

After he was inducted into service in the middle of the previous November, he entered active service on the last day of that November and was subsequently sent to the initial stages of his basic training in Kentucky, Pennsylvania, and eventually Georgia. For these first levels of his basic training, he had to leave home right after Thanksgiving heading into the Christmas season of 1942. Perhaps the single most important aspect about basic training, or boot camp, is that it was designed to completely reprogram youth from regular civilians into warriors. Accordingly, it aimed to place within the recruits a sense that they were expected to accomplish far more important things than could be expected from other young men still in their teenage years. This all occurred during one of the most intensely stressful periods of a young man's life, when he is intentionally kept isolated from contact with his family and friends and taught that everything he was before entering the Army was weak and lacked any real value unless and until he became a soldier in the U.S. Army. As a result, basic training was never an easy experience for anyone, as its aim was to break a young man down and strip him of his self-will.

The object was to ensure soldiers would obey commands from those higher in authority without question, without thoughts of self, family, loved ones, or fear of dying. By design this was extremely intense and aimed to touch every aspect of the person – physical, mental, and emotional. It all seemed much more surreal than Forest expected or imagined. He was miserable, and though he tried to work on his frame of mind, endeavoring to remind himself that this would pass in time, he came to a point of desperation. There was nothing that could help him it seemed, but he arrived at the conclusion that if he could only go home, even for a short time, he might be able to continue with the training. The truth of the matter was that he was desperate to work through this and it seemed that there was no place like home, and no one except his dad, who could help him. The problem was that during this phase of his training there was no going home for any reason, with the remotely possible exception of a severe family emergency. Reluctantly, driven by desperation, he wrote his dad and began to pour out his heart.

Back in Ohio Levi opened the envelope from his son and began to read the letter. Forest asked his dad if he could do him an enormous favor. Delighted yet surprised Levi thought about how great it was to hear from his first-born, and that he was a little surprised to receive the letter when he did. Based on the tone of the letter Levi wanted so badly to inquire if everything was alright, had anything bad happened, was Forest alright?

Barely holding himself together, Forest wrote that he did not believe he was alright, that he may not be able to get through the basic training. Forest's words made it clear that the training affected him in ways he could not have anticipated. In turn arose the need to be home for a bit. He had no thoughts of leaving the service, just going home for a little while. Levi recalled that Forest and his fellow troops were committed to complete all their training there before they would be allowed any leave. Furthermore, Forest confirmed this in his letter. Levi, feeling his heart starting to break for his son thought, "Well how would I possibly...," when Forest's written words interrupted his dad with "...family emergency, a severe family emergency is the only possible way for a man to return home on military leave." Levi knew there was no family emergency, and recognized his son was suggesting he contrive one. Levi thought to himself "Both of us could get into some serious trouble if we attempt this and get caught." Forest, his words heavier now, explained that he would not ask if he thought he could last. When Levi read, "Dad please, I beg you, write a letter to my command here explaining our urgent family emergency back home and request that I be allowed to return home for emergency leave. If you convince them then I might be given leave to come home. I do not feel right, concentration is difficult, sleep is fitful at best, and my stomach is always in knots...," his heart began to break. He felt himself submitting to his son's request. He had the camp address and sat down to address his letter to the commanding officer. Levi had a father's heart, his firstborn son, and eldest child, was in trouble, and it seemed only he could offer any hope. The more Forest poured out his heart, the more Levi melted inside and suddenly he could feel the pain in his son's words as though it were tangible. Maybe he relented more quickly because this was the only child of his late wife Mae, but several days later a letter arrived at Fort Benning, Georgia explaining the details of the dire emergency in the Birch household that had suddenly arisen and requesting that Forest be given leave to return home.

Forthwith the name "Birch, Forest S." appeared on the records giving Forest two weeks leave to go home. Perhaps it was just the magic of being home, talks with his dad, or a combination of all these things, but this break seemed to give Forest the resolve to finish what he had started. After the two weeks were up, he left southwest Ohio and returned to Georgia to finish his basic training, after which he was shipped to England for even more training. Now here he was, somewhere out on the North Atlantic Ocean, far away from home, farther than he had ever been or dreamed of being, headed to England where he would be even much further from home and without the chance to go home for any reason this time, unless the reason was severe injury or death. As he lay there wondering about what his training in England would be like the thought came to him, "I won't need to go home, will I? No, I won't because I have already dealt with this and my time at home during basic training has given me the strength to finish."

With this behind him he thought back on the completion of his training back in the United States. Training was never-ending. At the Replacement Training Center (RTC) as soon as one unit was ready, a group of soldiers was transferred out to form a new core for another unit, while other soldiers were promoted to fill the vacant slots and new groups of soldiers were brought in. In this way the unit kept up the training cycle until everyone was brought up to speed. The training period for a soldier arriving at the RTC averaged thirteen weeks. To complete this portion of his training Forest had to pass the Army Ground Forces test. This involved completing a minimum number of push-ups and pull-ups, a seventy-yard run while carrying another soldier of equal weight on his back, and a four-mile march in fifty minutes.

During this march, he was required to carry a full combat backpack containing toiletries, socks, and various other personal items. He wore a cartridge belt, onto which he hooked his first aid kit, a bayonet, a canteen with mess kit, and a collapsible shovel. He also wore his helmet and carried his ten-pound, 0.30 caliber semi-automatic Garand rifle. Added together this amounted to seventy-four extra pounds he would have to carry in the field. The unit to which he was assigned was infantry. The process by which he was assigned to his unit seemed simplistic to him, rather than the highly complex method he expected. A group of men arrived at the reception center, lined up in alphabetical order, and was divided into two groups. One group was sent to be trained as truck drivers while the second group was assigned to an infantry unit. This was by design, as the Army chose this way out of its concern to place the right number of men where they were most needed. By this process, however haphazard it may have looked, Forest became a private with a 57 mm anti-tank gun crew in the Headquarters Company of the 1st Battalion of the 175th Regiment of the 29th Infantry Division of the United States Army.

As he thought on all these things leading up to this point in his life, he thought again of home, and especially holidays at home. Maybe it was due to the timing of his induction into the service, but Christmas at home came to his mind. The Birch household was not a wealthy household by any means. In fact, they did well to get by, and there was no money to spend on gifts, even at Christmas. Each year Levi and his wife endeavored to make their Christmases unique and special in their own way. Thinking back on the Christmases at home over the years, Forest recalled getting up early in the morning and coming into the kitchen, which was a hub of family activity and the warmest room in the house during the winter months. There on the kitchen table the plates set out for the family members were all turned over lying upside down. After they were all seated and given permission, they turned their plates over. There, beneath each plate were two sticks of hard candy. If they happened to have had a good year, in addition to the two sticks of candy each one would find an orange sitting beside his or her plate. That was Christmas at the Birches.

Forest learned to appreciate what he had and hoped that whenever the time might come for him to have his own family, he would teach his children not to be unappreciative of what they had, and not to desire to live beyond their means. Rather, his hope was to make things better for his own children than they had been for him. Laying here in his own bunk he thought, "If I can live through this, go home, get married, and start my own family, I sure hope I can make an even better life for my kids, and assure they finish high school." Though it did not dawn on him at the time, his presence on this ship was the first and most important step he could take, and certainly the most self-sacrificing act, toward making things better for a potential future family.

He learned lessons from his own experiences at home, as well as from his dad's siblings, through their examples and stories he heard over the years, especially from one of his dad's brothers, Howard Birch. Uncle Howard could tell more stories without running out of things to share than anyone Forest ever knew, and Howard told true stories from his own life and daily routine. He was able to find humor mixed with a bit of mystery and drama in the life he lived, and therefore in the stories he told. Forest greatly admired the ability of his uncle to recount events and people from his daily life and weave these into stories that would keep you on the edge of your seat or laughing, trying not to fall out of it. Often when he visited with the family in Kentucky, he looked most forward to sitting on the front porch, having something to drink, and listening to Uncle Howard's accounts of daily life as a tobacco farmer.

Laying there, Forest thought he would give anything to have a book with all the stories, or even just a few of the stories Uncle Howard told the family over the years. How priceless and precious it would be to have something like that. People do not usually think to copy down the stories from daily life or to think that such stories are worth putting down for others to read. To have been able to read some of Howard's stories would have meant the world right then. Of course, it also occurred to him that reading one of Howard's stories was nothing like listening to him tell it. Maybe that was really the magic of it, being able to hear the stories in person. Howard told his stories with such delight and humor, and in his unique way he had the ability to pull you right into the story as though you could see the sights and hear the sounds. Uncle Howard was a master storyteller, and the best that Forest had ever known.

He thought of the unique qualities of each one of his uncles and other relatives. One of Howard's stories about his brother Lonnie came flooding into Forest's mind as he lay there moving back and forth with the ship. Lonnie was a small man in height as well as girth. Throughout his life Forest made first impressions of his relatives and thought Uncle Lonnie looked like a leprechaun out of uniform. He was short of stature and somewhat soft spoken, and another favorite uncle. Uncle Howard used to tell the story of when he and his brothers were younger and working on the farm. During one trip back from the fields they brought a wagon load of hay back to the barn. At one point they had to stop and let Lonnie jump down from the wagon to open a gate so they could go from one field to another. Lonnie opened the gate and the wagon pulled on through and stopped so he could close the gate and assure it was latched. Upon securing the gate Lonnie climbed back into the wagon. His duties now completed he returned to his spot in the back of the wagon, folded his hands and said, "Okay let's go." Before he could sit down the wagon started moving with a quick jolt and flipped Lonnie out and over the back of the wagon while his brothers looked on, trying to help Lonnie into the wagon without dropping him for laughing themselves. The first time Forest heard his Uncle Howard tell this story, he laughed so long and hard his side began to hurt. He could picture a little leprechaun being tossed out of the back of the farm wagon in a Kentucky tobacco farm field.

Laying there thinking, he lost track of time, but he did not feel quite so seasick now. His thoughts led him throughout various periods in his life so that, for what seemed like a long time, images of home and his preparation for military service flooded into his mind as he lay, timing the rocking motion of the RMS *Queen Mary*, smiling inside at Tippy's antics, recalling his dad's help when he did not believe he could continue on, recalling Uncle Howard's stories and his unique way of telling them, and picturing the family leprechaun flipping over backwards out of a wagon. Such thoughts helped him to neglect thinking, at least for a little while, about where he was traveling and why he was going.

2. St. Ives by the Sea

WHEN THE GIANT shuttle of the seas finally reached its destination in Gourock, Scotland, the U.S. troops on board still had one leg left to complete
their trip. This meant traveling by train from Gourock, Scotland southward
to St. Ives, Cornwall, a small fishing town on the tip of the southwestern
coast of England. Taking in the view of the quaint village gave no hint
whatsoever to Birch and the other troops of the purpose of their trip. If
anything, it helped, at first sight, to forget about what they had signed up
to do. What Birch saw before him was a picturesque, coastal village where
life seemed to ebb and flow with the tides of the sea upon which life there
depended. It was as if the world into which he stepped had been ongoing
with a sense of normalcy known only during peacetime.

Legend attributed the origin of St. Ives' name to the arrival of the
Irish saint Ia, of Cornwall, in the fifth century. The parish church still bore
her name and the name of the town was derived from it. Situated on the

western shore of St. Ives Bay, the harbor was sheltered by St. Ives Island. St. Ives Island was not actually an island at all, but rather a small grassy peninsula, called a headland, which connected to the mainland by way of a narrow isthmus. The narrow and uneven streets of the town appeared to gracefully weave their way among the places of business and residential dwellings. All the local structures were built with granite, metamorphic rock, slate, and other geological materials lying beneath the surface of this area of England known as Cornwall. Fishing had been the main source of income there since medieval times and later mining was added as both a source of income as well as a valuable resource for materials useful to everyday life. When the U.S. troops set foot on Cornwall ground, St. Ives was still one of the most important fishing ports on the north Cornish coast. The only factory in existence there made fishing nets. East of the businesses and winding streets a long pier ran south, parallel to the beach that faced St. Ives Bay. While the original construction date for the pier is not known, the first reference to the village having a pier dated back to 1478. In the years since, the pier was re-built sometime between 1766 and 1770, using a design that included an octagonal lookout. Over time the pier was lengthened. All this history occurred in this little coastal hamlet prior to the United States becoming an independent nation. This quiet little fishing village on the extreme southwestern tip of England became home for Birch and the men of his unit for the next year. Other than the abundance of U.S. soldiers, there were few indications in town of an ongoing world war.

Birch took further training there, remaining in England from May of 1943 until the first week in June of 1944. It was a scene that contrasted with the conditions under which he lived during his training in the United States. Here in St. Ives, he had the finest sleeping quarters and the best food the United States Army could offer, rather than bunks in a barrack and K-rations. For now, he had the best accommodations since entering the service, and even better than he lived with at home, and he did not hesitate to appreciate it. He had what he considered to be amazing access to a recreation area that one could not believe without seeing it. The United States Army took over a hotel in the town and lodged Birch's unit there. The building gave the soldiers a great view the beach, which Birch accessed by descending a very long set of steps, crossing the main street, and descending further down another long set of steps leading out onto one of the most beautiful beaches he thought anyone could imagine. During the day he and the other troops pretty much had the beach to themselves. At night though, the local civilians came out and joined the U.S. soldiers in enjoying the beach, ocean, games, and idle chatting. By all indications it was an English – American summer in St. Ives during 1943.

The town sat in one of Cornwall's many granite-intrusive, sedimentary-based valleys. All of this seemed strangely welcoming and in time Birch became attracted to St. Ives and enjoyed getting acquainted with some of its residents. Birch enjoyed his time in St. Ives, although he both knew that the day would come that he would have to go. He had no promise of being able to return, but he knew that given the opportunity he would like to visit St. Ives again. He had not been to the coast back in the States, never visited a beach, or experienced what it might be like to live a life by the sea. He was born in Huntington, West Virginia and then moved to Kentucky as a young child with his family. Later, in his teenage years, he and his family moved to southwest Ohio.

Now here he was, twenty years of age, soon to be twenty-one, thousands of miles from home and beginning to learn that the world was more than he ever knew. Birch's daily routine consisted of staying in a comfortable hotel, going down to the beach to take his exercises, play volleyball, and engage in what he referred to as "…this, that, and the other." The "this, that, and the other" included as many things as he could manage to squeeze in during his time there. He determined to take advantage of every opportunity to enjoy the unique opportunities of this hamlet by the sea. He particularly relished the experience of feeling the continual ocean breeze as it explored its way through the rooms of the hotel. It was a soft, pleasant whisper from the ocean that made him feel as though all would be well. Often, he closed his eyes and listened to the leaves clapping outside as they swayed to the rhythm of the salty zephyr carrying the aroma of saltwater, a new and welcome fragrance for him. What he saw, heard, and felt there that was so common to coastal towns around the world was a fresh and wonderful experience to him.

He knew down deep his attraction to St. Ives and Cornwall in general had to be put in perspective, as he was continually reminded of his reason for being there. He was there on a specific mission, not for leisure. This reminder came forcefully to mind by way of periodic visits from the U.S. Navy during their monthly visits. With the Navy's arrival, the soldiers loaded up their packs, along with everything they had, and boarded military trucks which took them to one of the shipyards somewhere else in England. Traveling in the trucks was not comfortable by any means. Birch was certain that he could feel every single bump and rise in the road, no matter how small or insignificant it might appear. Such routine trips by truck inevitably caused irritation to get the best of the soldiers. It was a release when one blurted out "Sarge can we not develop a way to travel in a truck without injuring or killing the troops before reaching our destination? For goodness sake, one would think this is the worst possible way to treat the ones doing the fighting." "Alright, pipe down," said the sergeant. I don't like this ride either, if you can call it that, any better than you do." Once they reached the designated shipyard, Birch and the other soldiers loaded onto ships and launched out onto the waters just as if they were leaving England. It really felt as though they were leaving England altogether, especially when they traveled so far out to sea so as not to be able to see land or anything else but water.

Out on the open sea every direction looked the same as the sunlight danced continually on the water everywhere they looked. Once at sea they maneuvered around for a couple of weeks and focused their training on the ability to get off the larger ships onto smaller boats by climbing the large cargo nets often seen hanging on the ships' sides. While this was far more difficult than it appeared, the difficulty grew immensely when large numbers of soldiers were all climbing on these large rope nets at the same time. As the wind blew across the sea, bobbing the ships up and down and rolling them back and forth in the water, Birch tried his best to make the climb. The cargo net swayed out and back in, constantly moving away from the ship and then back against it. It was not like a ladder whereon people climbed one at a time. In addition, he and the other soldiers on the netting were not Navy men with some experience in this sort of thing. Along with many others he had to learn to climb the hard way, which involved grabbing onto the vertical rope sections rather than the horizontal. This was contrary to the way he was used to climbing ladders back home. This ladder of rope could accommodate large numbers of men at any one time. With such crowds trying to maneuver up or down the net, grabbing the horizontal sections inevitably led to being stepped or stomped on by the soldier above. Many were heard yelling at their fellow soldiers, "Watch it will ya," and "Hey Mac that's not the net that's my hand, get off, off!" And on it went as each one learned the difficult lessons of maneuvering the cargo net. It was a configuration that enabled many men to climb on board or disembark, but there was a right way and wrong way to do so.

The men practicing now on the cargo nets were Army, and Army infantry at that. They were land soldiers learning to transfer from ship to ship by making mistakes regarding how to climb and not to climb. As a result, there were times when they lost men while making transfers between boats. At sea storms often developed, roughing up the waters and causing the Army to lose several men in a single instance. After successfully making the transfers to the smaller boats they began their simulated amphibious attack. For the time being they trained, and trained, and then trained even more. They did this over and over and over and over, until it became a dread whenever they received orders to pack up and leave their comforts in St. Ives.

One small welcome aspect of their training maneuvers was that they ate well while out at on maneuvers at sea, thanks to the U.S. Navy. After eating Navy food Birch was convinced that the Navy must have the best food of the armed service branches. It suddenly occurred to him one day that instead of serving food to other soldiers, as he did during his K. P. duty on the RMS *Queen Mary*, he was now being served by others, U.S. sailors nonetheless, and served food that he could brag about. Though he came to dread the grueling repetitive training routine, he and the others relished going out with the Navy on training maneuvers if only for the good eating provided. After engaging in this type of training over and over, he and the others in the 175th began to voice their musings about it all, "Why is it worth it?" Birch thought, "We have trained, trained, trained, and trained some more. We know it all by memory so what are we really accomplishing now? Anyone would think it has finally been drilled into us." Eventually they got to the point where they knew it all by rote, but the Army still grilled them and grilled them and drilled them and drilled them until they thought, "What is it gaining us...Why are we doing it?" Over time though, he and the others would come to see that such repetition, seemingly ad infinitum, paid off well. Before too long it became clear this was the reason why any of them were still alive in the days and weeks to come. While they all knew they would be part of the invasion forces, they did not know to what extent until about two weeks before it happened. For obvious security reasons they were not told anything until just before the invasion. For now, after they completed their training and sea maneuvers they traveled back to St. Ives and spent the time there just as they had been doing. That was really the only place they could wander about without a pass. They lived there together with the civilians in town, and sometimes in the evening Birch obtained a pass to go out beyond the town limits.

One evening as he and others in his regiment walked along the beach enjoying the ebb and flow of the waves, he fell silent. Thoughts began to form in his mind about their mission, what it might entail, where it might take him. He did not really know the mission details and could not talk about it to anyone even if he did. Still, he found it difficult to think about what his part in the war might be. Adding to his stress was that it became more difficult to keep his questions to himself without being able to talk aloud about them with others in his regiment. None of the young soldiers really wanted to talk about the impending time of departure when the time came for them to leave, with no assurance of being able to go home. Instead, they did their best to enjoy the time by the sea, leaving all mission-related conversations alone. As they walked along the beach, he thought about what it might be like when the time came for actual fighting. The one person he really wanted to talk to about it was his dad, Levi. He had talked to Levi when he was given permission to go home on emergency leave, but now it was different. His training here off the shores of England revealed much more about what kind of conditions he would be facing. Not only that, but he realized and personally experienced the inherent difficulties in the effort to transfer from a larger ship to a smaller, more maneuverable one which would be taking he and his fellow American soldiers onto some beach where the enemy would be dug in and waiting. Levi was still his confidant and yet he did not want to worry his dad by speculating about dangers he might face prior to fighting the enemy. For now, he decided to enjoy his time in St. Ives by the sea with his brothers in arms.

Beginning in early January of 1944 and continuing through the first week in February Private Birch and his unit began more intensive training to simulate the conditions under which they would be functioning after leaving behind their comforts in St. Ives for the final time. This leg of the training exercises was called "DUCK 1." As the first leg of the major exercises, DUCK 1 was likely the most important, for any mistakes and flaws that might show up during its execution could greatly affect training and planning for the exercises to follow. It was during this time, on February 3rd of 1944, that Private Forest Birch was promoted to Private first class (Pfc.) Forest Birch. He and his crew soon found out that training maneuvers were by no means simple dress rehearsals. Instead, they could be much more complicated and hazardous than one could imagine. This was evidenced by the lives lost during the training exercises where the Army infantry had to learn to maneuver the cargo nets when transferring to and from the smaller ships, called Landing Ship Tanks or LSTs. As a part of DUCK 1 Birch and his crew launched out to sea just as they had done so many times before and would do during the actual invasion. The ships loaded with soldiers went out well past being able to see any land then circled back for a series of simulated landings. The soldiers practiced their transfers from ship to LST and then moved in toward a part of the southern coast of England known as the Devon coast. This area was made to resemble an actual war zone, much like what they would face on the day of the invasion. The secluded beach was named "Slapton Sands," which now represented their target for the actual invasion, the location of which still had not been revealed to them. All the U.S. soldiers knew for certain at this point, whatever their suspicions might have been, was that they were to conduct an amphibious landing onto some beach or beaches in occupied territory and take the positions being held by the German military. For now, Slapton Sands was that beach.

The beach had been converted to simulate a war-zone beachhead with a maze of mines, barbed wire, and concrete obstacles. In this section of England, the British government previously selected the area for conducting Allied training exercises and ordered the evacuation of all the homes and villages. The area around Slapton Sands in southern Devon was chosen due to its characteristics, which were thought to be like the target area where Allied forces would be landing. This was done because the U.S. armed forces requested that they be given a location in the U.K.

to practice the amphibious operations they would soon be conducting for real. These training exercises were designed to be as realistic as possible with respect to the type of work each unit was assigned as well as the actual conditions under which they would be working. Efforts were made to provide the soldiers with the actual sense of the chaos of battle. To help accomplish this, the Royal Navy, on cue, shelled the beach with live rounds of ammunition until just moments before the U.S. soldiers moved toward the beach, landed, and attacked it. This was the training Birch and his crew repeated so many times in preparation for the actual invasion. Just as they saw with the cargo nets, unintended results often resulted in the loss of lives. This was due to such things as miscommunication, mistakes, and even real encounters with enemy forces.

A sobering example was burned into Birch's mind as he kept abreast of the developments of other units participating along with his own in the training exercises at Slapton Sands. The training exercises on the Devon coast occurred over time in three major parts, 1) three DUCK exercises, 2) a FOX exercise, and 3) two dress rehearsal exercises called FABIUS and TIGER. Each of these parts was designed for specific units who had their own unique mission assignments to complete. In late April of 1944 approximately 23,000 U.S. soldiers assembled at various staging areas around England. This occurred under the strictest secrecy as the third leg of the training and was tagged "Exercise Tiger." On the first day the primary attacking wave was plagued by delays and miscommunication. In addition to this, a scheduling foul-up resulted in having a few of the U.S. small landing craft, called Higgins boats, which were full of U.S. troops, arrive at the beach at the same time the Royal Navy was shelling it. While the shelling was immediately called off, the participating U.S. forces suffered several friendly fire casualties.[1] The U.S. officers in charge of this rehearsal were livid, calling into the radio "Stand down! Stand down! Stand down!" Out of frustration they blamed the Royal Navy's mistake as though they were intentional. While the U.S. command was livid, disappointed, sad, and fearful of not being prepared, they could only hope for a better, more orderly, landing on the next day.

During the second attempt, an attacking wave of engineers and back-up troops approached the coast in a convoy of eight landing ships, the LSTs. To this point during the series of exercises there was no fear of encountering Germans, even though they patrolled the English Channel. This time as the LSTs headed through the entrance to the bay they caught the attention of nine German torpedo boats, called "*Schnellboote*" or "fast boats." The Allies referred to these torpedo boats as "E-boats." These were small, quick attack boats outfitted with torpedoes and 40 mm guns. Upon discovering an Allied fleet in the region, the E-boats immediately sped to intercept and attack it. Early in the morning, around 0130, the E-boats intercepted the flotilla and launched their attack. As eight U.S. LSTs made their way toward the coast, their crews were suddenly startled by thunderous eruption of gunfire and flashes of tracer rounds in the night sky. It was as if they had wandered into the war then and there. Orders among the LSTs were quickly issued to ward off the attack, but the fleet was entirely caught off guard. All they could do was try to get out of this as best as they could.

While British forces were monitoring the approach of the E-boats toward the LSTs, due to another error the British and the U.S. Army were operating on different radio frequencies. Making matters worse was the fact that the LSTs' main escort, a Royal Navy destroyer called the Scimitar, had sustained damage earlier on in the evening of the previous day and had returned to port for repairs. When the shooting started, the U.S. soldiers' only protection ended up being a small 200-foot-long warship, called a corvette, named the Azalea. Within about thirty minutes after the commencement of the attack a German torpedo rammed into the side of an American LST (LST-507) turning confusion into panic. Soldiers on board heard a horrendous, thunderous explosion accompanied by sounds of crunching metal. Choking dust was everywhere as the explosion set the LST on fire, killed dozens of troops, and forced others to abandon ship. Men jumped over the side of the LST into the near freezing water approximately a twenty-five-foot drop from the top deck. At the same time the surface of the sea was on fire, fed by the oil slick from the damaged LST. While LST-507 burned, another

LST (LST-531) was struck in rapid succession by two torpedoes and erupted into a ball of flames. Its crew, or what was left of them, had no choice but to hurl themselves overboard. During all this a fourth torpedo plowed into a third LST (LST-289) and destroyed its stern. While LST-289 managed to stay afloat despite the stern damage, both LST-507 and LST-531 sank within a few minutes after being hit.

Survivors of the attack either huddled together in life rafts or floated in the cold water. Since many had not received proper instruction on the use of their life-jackets, they drowned under the weight of their personal combat gear. During the attack, the Allied fleet scattered but later returned toward the coast. Once it was known that the German E-boats retreated, a lone LST and a Royal Navy destroyer returned and began pulling survivors out of the water. By then hundreds had either drowned or succumbed to hypothermia. Rescuers saw a mass of dead soldiers bobbing in the bay. Trying not to think too much about it lest succumbing to despair, they worked to collect the bodies and stack them on the deck.[1]

The process seemed very methodical and impersonal to those looking on trying to resist the tendency to fixate on what occurred, as that could be more than one could bear. Witnessing these mishaps that cost so many lives and knowing some of those soldiers involved made it that much more difficult to continue the training in preparation for the actual invasion. The soldiers of the 175th Regiment had been made to expect that their landing on the actual day of the invasion would be routine, perhaps not very much different than their training at Slapton Sands earlier on, that is, before everything went wrong during Exercise Tiger. Given this, Birch wondered, "Exercise Tiger included preparations and efforts to control conditions as much as possible and still things went very badly. If this was a dress rehearsal, what in the world is the real thing going to be like?"

It would not take him long to find out. The day soon arrived for final preparations for the real thing. When the call came it set in motion a domino effect of calls going out to all the pubs and other places where the soldiers might visit, for all soldiers to report back to their quarters. That was first. Second, they were told not to tell anybody anything. After going back to their quarters, they were told, "Load up, pack up everything." After they loaded up their belongings they were directed to a back alley where they found trucks gassed up, lined up, and ready to go. There were no Army vehicles on the streets of the village, and for anyone walking the streets of St. Ives nothing looked much out of the ordinary. One would never know anything was in the works. All the activity and vehicles to take the soldiers out of town for good were crowded into the alleys in the back of the businesses lining the town's streets. According to a plan that had yet to be communicated in detail to the soldiers, they were to load up and ship out of town before the townspeople could know they had left. At least that was the plan. No one was supposed to know until after they left, but somehow word got out. Some of the soldiers had girlfriends, which meant that somehow, some way, they took the time to get word to their girlfriends through anybody they could manage to get to help them. Of course, the result of this was that there were gathering crowds of people in the back alley wondering what was going on and what it meant for them. Though in appearance St. Ives seemed insulated from the events transpiring around it, the truth was that everyone living there felt the heaviness and uncertainty of the war. No one had been left untouched. Birch and his buddies did not know much more about what was going on than did those people from town. He had no opportunity to call home, but he wished so much for his family to know this was the time. Thinking about this he noticed a tear running down his cheek, and recalled that his job was to stay alive, and to stay alive meant he must remain focused on what he was trained to do. As the soldiers loaded up and the military convoy pulled away and headed out of town that was the last that he would see of St. Ives, his home for a year. It was just a small hamlet that he came to love and appreciate more than he expected or could express, over 3700 miles from his home in the United States.

The truck convoy made its way through the English countryside like a giant green snake curving its way forward on its way to its destination. Birch's experience was the same as the other soldiers in that they shipped out and were all told nothing until they finally reached their planned destination, which happened this time to be out on the cold, barren moors of England. There they established a "tent city" out in the country where there were no visible signs of homes, people, or anything else except a lonely roadway. It was distinctively unlike their comfortable quarters in St. Ives and was a very desolate looking land. It gave Birch the sensation that he left England altogether and ended up in some far away land where civilization had yet to touch. After the disasters of their training at sea the thought occurred to him that such a far-away land beyond the reach of civilization might be a good change. There in the desolate land his unit lodged in tents and were told they would be briefed on their mission for the invasion. From the time they were informed about this briefing they were not allowed to speak to a civilian if they happened to come across one. As it turns out there were people and civilization nearer to them than it first seemed. An occasional civilian could be seen using the single road running through the area, but if anyone happened to pass that way the troops were not allowed to speak to him or her. To make the situation more surreal, they were placed under guard by their own fellow U.S. soldiers. Another outfit was assigned to place Birch's unit under guard. If they went to a movie, they went with a guard. When they went to eat, they went with a guard. As they slept, they slept under guard. In the end he personally never actually spoke to a civilian from the time he left St. Ives until about three weeks later when he and his unit were finally able to regroup after the invasion.

While lodging on the moors the unit waited to be briefed on what they were chosen and trained to do. A couple of days prior to June 6th they loaded onto ships and cruised up and down the English coast waiting for the order to go from Supreme Allied Commander Dwight D. Eisenhower. Their destination for the invasion was to be Normandy, France. The 175th Regiment was to go in on Omaha Beach. All the thousands of ships that were to participate in the invasion came from many different ports of England. Some ships loaded in one location and some in another, some loaded earlier than others, but eventually they all gathered to coordinate their invasion of Normandy. They knew the area of the English Channel that they were going to be crossing was the widest spot between Great Britain and France. The hope was to be able to use an element of surprise against the Germans, who were expecting an Allied invasion at the narrowest point in the channel, in the Strait of Dover which ran between Dover, England and Calais, France. Everybody was on edge, floating up and down the coast waiting for orders. All their training had them poised and ready to go, but now they had something else, one last thing, which they needed to learn – how to wait and then wait some more. Waiting was not something for which they were able to adequately train.

They finally received word that it was going to happen on June 5th, but stormy weather held it over another night. Eisenhower made the decision to go on June 6th. The 175th did not know the exact hour but they knew it was right there before them, and that they were among the most trained and prepared soldiers on Earth. It would soon become clear that the repetition of their training would enable many of them to perform automatically, without taking too much time to think it through. Taking time to try considering everything could cost one his life and Birch was beginning to see this evidenced more and more over time. All units kept in contact with the main headquarters and the 175th finally received word that they would be in the third wave. One wave of ships was to take in the engineers and those who could help clear a way for the fighting men to get onto the beach and eventually up and over onto higher ground.

Things were becoming more intense and real. It was a little easier for Birch to forget about what he was facing when on the RMS *Queen Mary* thinking of home, and later in St. Ives enjoying the town, the beach, and the civilian company in between his training at sea. Now he could think of nothing else but his mission as a part of the 175th Regiment, along with the dangers awaiting him. Though he felt pride in the 175th, he was unaware of its long history. One of the oldest regiments in the U.S. Army, it was first organized in 1774 under the name "Baltimore Independent Cadets." [2] In 1776 it was absorbed into a Maryland Battalion and fought heroically at Long Island. Eventually this unit became seven different Maryland regiments collectively called "The Maryland Line." Several years after the Revolutionary

War the Maryland Militia was reorganized. At this time several volunteer militia companies of Baltimore, including former members of the Maryland Line, became the 5th Regiment of Maryland. At the beginning of the Civil War the 5th Maryland broke up, many of its members going south to fight for the Confederacy in the 1st Maryland, C.S.A. Two years after the end of the Civil War the 5th was reactivated. During the First World War in 1917 the unit became known as the "Dandy 5th," was federalized, and designated as one of three infantry regiments attached to the 29th "Blue and Gray" Infantry Division, which had just been formed in 1917. Prior to being transported overseas, the 5th was consolidated with two other Maryland infantry regiments to form the 115th Infantry, which fought in a major Allied offensive campaign just prior to the close of the First World War. After the First World War the 5th Maryland was again reorganized. Subsequently, prior to activation of the 5th in February of 1941, the U.S. War Department assigned it a new designation: the "175th Regiment."[2] This new designation was considered necessary to avoid confusion with the regular Army's 5th Infantry Regiment.

Whether or not soldiers of the 175th realized their unit's long and distinguished history, they were now a part of it. Standing on the deck of the ship, their training complete, they were gearing up for an invasion into Europe. The puzzle pieces had finally fallen into place as Birch thought back on his training with the 57 mm anti-tank gun back in Georgia, his subsequent training in England shipping out to sea, and then simulating amphibious beach landings. Now there was one enormous difference - the target in England had been vacated and no one was firing back. This was one aspect of the mission that could never be simulated in a manner to make it real enough for him and the others to adequately prepare. The infantry would eventually endure the hardships and horrors of war for which no training could possibly prepare them.

In a real sense the infantry was the military workhorse of the armed services. Not only was it tasked with fighting the enemy on the front lines. It was often called upon to do much more, including such operations as transporting food, clothing, weapons, medicine, and other supplies needed to continue the fighting necessary to win the war. In this sense the infantry's experience of the war was unlike that of anyone else. Pfc. Birch was infantry, about to step off his LST to experience, with thousands of others, something many of them would not live to talk about, and that the rest would not find easy to recall, let alone discuss. Everything up until now had been part of his journey. The moment he stepped off the LST into the waters off Omaha Beach, he would be at his destination. From here on lay a strange new world where the normalcy of life would seem lost forever.

3. Disappearance at Omaha

THE LANDING SHIP TANKS (LSTs) that carried them to the beaches were unique vessels capable of carrying troops, vehicles, and equipment. Pfc. Birch waited with his unit on LST-262. He had never seen a U.S. Navy vessel quite like the LST before, let alone been on one. It was designed by the Allies as a wartime vessel designed specifically for conducting amphibious operations during the Second World War. At that time during the conflict the LST was the largest "beachable" landing ship making up part of the amphibious forces, which had only recently been established. This unique ship design supported amphibious operations by its ability to carry tanks, other types of vehicles, supplies, cargo, smaller landing craft, and especially large numbers of troops directly onto enemy beaches, without any need for docks or piers to be built beforehand.

On his LST Birch counted four decks, which included the lowest deck, deck four, where the engine and mechanical rooms were housed. Deck three was a cavernous area able to accommodate tanks, trucks, jeeps, associated equipment, and fighting troops being ferried to the beach for direct assault. Housed on the second deck were kitchen and dining areas, sleeping areas for the crew, and the captain's quarters. The top deck was where the bridge, communications room, and weapons were located. It was undoubtedly the specially designed hull with a flat bottom that gave the LST the ability to land on virtually any type of shore with a gradually sloping beach.[1] The ship was also equipped with a unique system of ballast tanks that could be configured for travel at sea or landing on the shore. The operation involved pumping the tanks full of water for trips out to sea and pumping them empty for landing on shore. As the troops cruised the sea earlier on in anticipation of the call to go, they learned the lesson of waiting. Now they were putting this into practice again, waiting to go.

The men of the 175th Regiment were full of anticipation as they awaited their turn to leave the protection of their LST to begin their work on the shores of France. None of them anticipated the sight that they were looking at now though. The first thing they noticed was the sky, which seemed dark from the amount and variety of aircraft in the air. In addition to the airplanes, the LSTs and most of the other ships were accompanied by strange looking airframes that at first glance looked like they were from another world, and certainly out of place in an amphibious invasion by armed forces. Floating over the ships were aerial balloons, called "barrage balloons," which were oblong-shaped silver balloons made of two-ply cotton embedded with synthetic rubber, and attached to the ships by long steel cables. The top sections of these balloons were filled with helium or in some cases hydrogen. Most of the ships, including the LSTs, were each equipped with a barrage balloon.

As he gazed at the dark sky and then at the cable attached to LST-262, Birch wondered about the purpose of this "mini blimp" hovering like a giant bird above the deck of the LST. It was clear that the purpose of these aerial craft was not transportation because they were tethered to the ship and had no accommodations for holding personnel or maneuvering. Spotting one of the sailors working on the LST he queried the nearest sailor about those mini blimps floating above. The sailor, on his way to his next task, explained these were for protection. At any moment the invasion fleet could face an aerial attack by dive bombers, which are nearly impossible to shoot down using anti-aircraft weapons. The low flying planes were certainly too fast for the gunners to have any real chance at them. The barrage balloons were affixed to the ship and floated up to an altitude of between nine thousand and twelve thousand feet. The long thin steel anchor cables holding the balloons were not visibly apparent to planes flying through the area. These cables had the ability to catch on to the wings of the low flying aircraft and cause a plane to stall, or even rip a wing off. Therefore, enemy pilots who were determined to attack the fleet were forced by the barrage balloons to fly at higher altitudes where they were much more vulnerable to anti-aircraft gunfire from below.[2]

These balloons were not only used on the ships but were also taken on shore previously by the 320th Barrage Balloon Battalion, which was the only African American battalion to storm the beaches of Normandy on June 6, 1944. Any area on the shore that was subject to attack by aircraft above was protected by attaching these barrage balloons to concrete anchors on the ground. The balloons themselves, as well as their cables, could destroy aircraft, especially at night.

Pfc. Birch looked again at his watch. It was well past their designated time to strike. The men of the 175th Regiment were learning to wait, which would prove to be a valuable lesson once they were on shore making their way inland with the enemy nearby. They were men bound together not merely by things such as age, ethnicities, or even the fact that they served together in the same unit. Rather they were bound together by "right-of-passage," the heritage which entitled them to the rights and privileges that were theirs by virtue of being Americans, whether they were naturally born U.S. citizens or came to the U.S. and became naturalized citizens. It was their time, they were waiting out at sea on the landing ships, they were ready and poised to strike, they had trained until their reactions were automatic, they had been provided everything needed for their mission, and they were waiting for the right moment. Everything was now in place, and they were here, on the very threshold of the war. It was the calm before the storm, and they were ready to strike.

The time was approximately 1330 local time in Normandy, France on June 7th 1944 and Omaha Beach was still not completely secured. The men who were part of Birch's anti-tank gun crew were waiting together as a unit out in the English Channel about two miles from Omaha Beach. They were supposed to have gone in on Omaha at 0930 but were still waiting on the LST. The instructions were that the 29th Infantry Division would hit Omaha Beach at a designated area known as "Dog Green." The fleet was equipped with mine destroyers and engineers, along with all the troops who had trained to go in ahead of, or with, the fighting troops to clear a way for equipment to get on shore. The resistance which they encountered was much higher than expected because there were more German troops than anticipated. Birch felt as though everything was suddenly moving very fast, leaving him trying to keep up with events. He wondered if the feeling would subside, and when he would become accustomed to the circumstances so he could anticipate problems and better prepare for them. For now, he felt like he was "running to catch the bus." This was not the ideal situation for him or any other soldier.

His watch had already revealed their plans did not work out like they all hoped. The familiar feeling experienced during the training mishaps that ended in disaster at Slapton Sands returned. Everything had been planned to occur in waves. The first wave included the engineers, mine destroyers, and crews who were to make roads for the vehicles and equipment to get off the beach once they got on shore. The main thing was to get the tanks, other vehicles, and different types of equipment off the ships and onto land. Nothing was going to work unless and until the troops had everything they needed. Though these operations were supposed to happen earlier, the holdups from the previous day caused delays, so that they did not happen until much later in the day. The delays causing his unit to wait for so long into the day occurred because the first waves of soldiers trying to help establish a foothold on Omaha were pinned down. As the men on the LSTs looked on, it seemed as though the Germans were killing their brothers-in-arms on the beach just as quickly as the soldiers hit the shore.

Pfc. Forest Birch and his 57 mm anti-tank gun crew were in the third wave. The earlier waves of fighting troops, and troops taking the necessary equipment to shore, were ahead of them. The efforts of the men of the infantry who helped with the fighting to establish the first footholds on Omaha Beach, combined with those men who charged the shore under fire to assure that the troops following behind would have what they needed, were crucial to the success of the landings. The men of his crew awaited the green light to leave their LST.

U.S. anti-tank gun crews typically consisted of ten men with a 57 mm gun towed by a truck, eleven men total. There were three such guns and associated crews in a platoon. Birch and his crew had orders to get on shore and set up their equipment. They trained to work with the other gun crews to set up their guns in a triangular formation, the three guns forming the points of the triangle with the intent to better assure taking out a German tank. If one gun could capture the attention of the tank, then perhaps one of the other guns could possibly disable or destroy it. The major problem with taking on a German Panther or Tiger tank was that it could not be immobilized from the front because the armor was too thick. Only a direct shot from the rear, or possibly from the side of the tank, would do the trick.

Their first mission was to make it to shore and unload the men, vehicles, and equipment. Then they had to make it across the beach to the road being prepared for them. Finally, they had to successfully climb the very steep grade that ran along the beach like the giant backbone of some half-buried behemoth. The climb there felt as though it were almost straight up, and the hill felt like more like a mountain to those who actually had to climb it. At one place along this ridge the slope came down to a v-shaped pass, the middle of which met the beach. At least this is what little Birch was able to make out, for he was not looking at any scenery that day. He was trying to be alert to everything occurring around him and it suddenly dawned on him that once ashore he would have to be aware of everything in a 360-degree circle around him. He remembered never to assume that everything is secure behind you. There in the small pass of the slope it was rough, but that was the spot where the engineering crews eventually made the road for the 175th Regiment and other units to make it up and over to the top of the cliffs above Omaha. During the battle to establish their foothold, one American soldier had managed to take eight or ten men, locate the low spot in the ridge, and start building a road to make it possible to get the awaiting men, vehicles, and equipment quickly off the beach and up on top of the hill. A little nervous as the time for disembarking got closer, he asked his sergeant if he should check the equipment again. Sensing the nervousness in Birch's question the sergeant told him not to take the time to do rechecks, because about the time he would start they would likely get the green light to go. Birch wanted to be sure he and the others were ready and had not overlooked anything. He stopped thinking and told himself to just calm down a bit. He knew what he needed to know, had trained on the equipment, ridden in tanks to get a feel for what the enemy feels and sees, and learned all the safety protocols for those times when things go south. "Trust your training, but most of all trust yourself," he told himself. "Right," he thought to himself, "Right, that's easy when you say it, but not so easy to do."

Up on the top of the cliffs of Omaha was a small hamlet named Vierville-sur-Mer, where very bitter fighting occurred even as the U.S. troops were still working to get off the beach. Prior to beginning the unloading process, Birch and the others could see some of the fighting occurring from their vantage point on LST-262 loaded with tons and tons of trucks, other vehicles, and equipment. One of the vehicles was the truck carrying his crew. It was a two-ton truck equipped with bench seats on either side in the back for carrying the men. His crew of ten men along with the driver and their truck would be going on shore with fifty-five gallons of gasoline mounted onto the sides of the truck, plus all their ammunition, which consisted of very heavy shells approximately eighteen inches long and about three to four inches in diameter. Most of the shells were high explosive (HE) shells designed to go through a tank. These were called Armor Piercing High Explosives (APHE) and were designed to pierce the hull of a tank then explode on the inside. That is the type of explosive materials they were carrying with them, and that made them a moving bomb should they and their truck be hit by enemy fire.

In addition to all of this, the LST carried a large semi-truck in the large, cavernous third deck of the ship. This semi-truck carried a large load of steel for building a bridge. The LST was unique in that it was designed to open at one end and let down a ramp directly onto the beach for offloading vehicles, personnel, and equipment. After they eventually reached the sands of Omaha the men had to wait until they received the order to offload. Offloading required that a tank unload first, then the truck with the bridging material, and then everything on the third deck would leave the LST in succession and follow the original tracks made by the tank. Since the tank had been loaded closest to the ramp end of the boat it had to leave the LST before any of the other vehicles. As it did so it made a path in the sand with its tracks. Every time a vehicle or piece of equipment went off an LST, it sank down deep into the sand on the beach, packing it down to form a hardened path.

Once the operation began it could not stop, and those driving the vehicles knew they would have to push the machines to crawl out of the ship, down the ramp, and up onto the beach. Birch stood on deck with the others, shells falling beside them, and did not know when it would be their turn, that is, when they would be hit by the shelling. He looked to either side of the ship and could see ships on fire and wondered what was going to happen. "We all look and act like kids scared to death. Do the Germans feel like this?" Birch thought, "Is it really the case that the various countries involved in this war have sent their kids, their scared kids, to fight a war that could determine the fate of nations?" It seemed they were unable to get consistent information. Instead, various conflicting information seemed to come flooding in. They heard the order "Get ready we're going in" but at the same time there seemed to be no movement to disembark from the ship.

When it finally came time for offloading soldiers and equipment, the first crew to follow the tank was a crew from another battalion of the 175th Regiment.[3] It felt surreal, watching a crew not unlike his, almost as though Birch was watching his own crew. While he looked on, the men loaded up in the back of their truck exactly as they had been trained to do. Then as the truck full of men and equipment made its way down the ramp, the driver carefully guided the vehicle off the LST. He lumbered along, headed straight down out of the LST so that he could drive directly to the top of the hill by way of the road that had been cleared for them earlier. When the truck tires moved off the ramp, the driver steered carefully to stay within the deep ruts made minutes before by the tank that drove off first, cutting a path into the wet beach sand. It had gotten very deep there and all the soldiers unloading did so under the assumption that the order to disembark would not have been given had the beach not been previously cleared of mines. Still, they followed the safest procedure, which was to stay inside of the tank's ruts in the event something had been overlooked. By doing this they should be able to avoid any live mines that might possibly still be on the beach. All of this happened quickly, and Birch had no time to assess all the possible scenarios. Instead, he watched the truck as it followed the presumed safest path onto the beach.

Unknown to the driver and men on the truck was that not all mines on Omaha Beach had been cleared. It is quite possible this may not have been a problem for the truck had it preceded the tank. The mine hidden beneath one of the tank's ruts apparently was so deep that the tank pressed down the loose sand, compacting it to the point where the mine could easily be triggered with pressure from another vehicle. As the truck followed the tank and moved off the LST ramp its tires pressed into the tank's ruts and in the process of doing so triggered the mine. Next in line, Birch and his men watched in shock and horror as the truck, men, and load of bridging steel disappeared right before their eyes. They were standing ready to go off, and it all just disappeared in an instant. In one sudden explosive burst there was nothing left. The men watching went berserk, with different voices shouting and crying out, asking what happened, how it could happen. The effect on them was immediately apparent in their eyes, on their faces, even in how they sat or stood waiting for their own turn to ride down the same ramp. After all they were to be the next ones to exit the LST.

In a bursting, deafening flash, Birch was impacted by a shock wave that he did not even feel or notice, but he felt something else flow over and through him, pervading his entire being which he could only later identify as unadulterated fear. It caused him to feel cold and hot at the same time, as he broke out into cold, clammy perspiration all over his body. Then he felt it in his limbs, as though strength literally flowed out of his legs. It was as though something tangible deactivated him and rendered him helpless. It was not an experience he had ever exactly known, and it certainly was worse than the dark sense of desperation that he felt back in the States during his training in Georgia. Birch mumbled to himself, "How in the world did this happen? Our first landings just started, and our troops have been working to clear the beaches."

All that remained after the explosion was a little fire left over to the side. Then, as though some time had passed, a couple of the men from the disappearing truck dropped back to the ground, on fire. He could not really tell what they looked like because they were black and burnt like charred meat rather than men. Still soldiers and crew members of LST-262 ran, grabbed the bodies, and took them to the hospital on board the LST. That was literally all that was left as far as he could tell. When his senses eventually started to return to him the first one to begin to clear was his sense of hearing and what he heard was shouting. The commander of his battalion was standing and yelling, "Get moving! Get moving! Move, move, move! Get off this boat!" Omaha was obviously not yet entirely secured, the Germans were shooting at them, and in the midst of this they still had to unload the LST. Their commander was charged with getting all his men and equipment off the ship, only now he made sure that the rest of the soldiers walked beside their trucks. They were told, "Men do not, I repeat, do not load up into the truck. Just be sure to stay with it and walk beside it."

This was one of Birch's moments of truth, for he had not yet left the confines of LST-262 and yet had already witnessed the ravages of war that caused a truck load of men and equipment to disappear right before his eyes. Now the time had arrived for him to get off the LST, placing him in the difficult position where he had to force himself to get off the ship by way of the same path as the truck and men that had just been obliterated, thereby leaving behind the safety of the landing ship. As he started to move Birch thought, "Perhaps this was just a fluke and the mine that the tank failed to trigger was the only mine…it just happened to be. That bridge truck that sank down so far and set the lone mine off; it was just one of those things." Still the crew of that truck was gone in an instant. Birch made sure to walk beside his truck. As he did, he could not help thinking, "Welcome to Omaha Beach."

As his unit moved off the LST toward the first destination, they were accompanied by communication troops whose job was to string phone lines for maintaining contact with headquarters. Along with the phone equipment, they carried jeep-mounted radio equipment with them everywhere. Wherever he went Birch could see phone lines laid out like dark strings of spaghetti across the ground, running on continually as though they had no end. With all this effort and assistance, for all the units involved, they still had not managed to get to shore until around 1530 hours. That was Omaha Beach. Not only were things much worse than he and the other members of the 175th had been led to believe, but now the men unloading from LST-262 were discovering that they were off target of their landing spot on Omaha Beach. They ended up unloading the ship over a mile east of where they were supposed to land, the area designed as "Dog Green" which was the draw just below Vierville.

Instead, they had been taken by the Navy to another draw further east along the coast, to an area named Les Moulins, which according to the Operation Neptune map was referred to as "Easy Green." Operation Neptune was the name of the landing portion of the Allied invasion, known overall as Operation Overlord. Their sergeant was livid and verbally expressed his ire over why the Navy could not get anything right when it comes to navigation? After all, navigation was their job; it was what they did as part of their routine. As it turned out, though, their drop off at Easy Green was by design on the part of the U.S. Navy. The Navy believed there were too many mines at Dog Green that had not been cleared, and that landing at the Les Moulins draw would be a safer option for the troops. Once all of this was explained to the sergeant, he was a little more understanding. It was a sad, tragic irony that after this effort to avoid live mines, the first truck to move off LST-262 at the Les Moulins draw was destroyed by a live mine that was somehow overlooked. Landing on shore at Easy Green meant that Birch and the others in his unit would have to hike west for over a mile back along Omaha Beach toward Dog Green, and then start their mission on land from the position they originally intended. Moreover, they had to make this trek by foot while still under intermittent fire from machine guns, hidden snipers, and artillery fire from fortified positions. They later learned that many of the enemy machine guns were "manned" by women. The Germans adopted the habit of forcing some of the French women to man the machine guns when they moved out. Then they moved further back and set up heavier defenses to catch the Allies off guard. Once his crew reached their original designated area, they still had the steep climb ahead of them up to Vierville where heavy fighting was still occurring.

Added to all of this was the fact that they had to take care as they made their way along Omaha Beach under fire and walking hunkered down as much as possible. What they encountered took them off guard, for there on the beach, in the sand and water, they had to step over and around the bodies of their fellow U.S. soldiers who were killed during the very first landings at Omaha. Looking at their bodies, the men of the 175th spotted on many of their uniforms the distinct, recognizable blue and gray circular insignia that was in the form of a patch for the 29th Infantry Division, of which their own 175th Regiment was a part. They knew they were on consecrated ground, and their march became a silent, soggy trek across the temporary graveyard formed by the surf pushing sand up and over the bodies of the men laying there on the beach. During this trek Birch knew the others were doing the same thing that he was at that moment, promising those lying there that he would make sure their deaths were not in vain. He would pick up where they had to leave off. As he thought about his promise to them, he noticed his breathing cycle aligned exactly with the cadence of his steps. As he made this solemn promise the only sound heard in addition to the intermittent weapons fire, was the pounding surf as it played a haunting dirge for the fallen men during the battle on Omaha during the past two days.

For Birch the war was just getting started, and already he was made to feel a deep sense of dread for what lay ahead. If they had been this wrong about the conditions under which they would be landing what did that mean about the sources of information so crucial for them to succeed in their missions? If the landing and effort to get to their original assembly area had been this difficult and discouraging, what did that say for the remainder of their tour of duty?

Along the draw at Vierville, Allied equipment was lined up

bumper-to-bumper waiting to get through, waiting for the infantry men and the machine gun crews to get in and clean out the enemy opposition entrenched there. The fighting on top of the cliffs over Omaha Beach around Vierville had gotten so bad that Birch and his crew eventually had to get out of their vehicle, leave their truck and anti-tank gun behind, and start fighting individually with their personal firearms with the aim to help clear out the opposition and secure the town. Before darkness came, they were able to cut down opposing fire enough that it "quieted down" to the point where sporadic gunfire could still be heard. At this point it was deemed safe enough for them to go on through, but this still had to be done by walking, watching, and hunkering down.

He and the others in his crew, and in their regiment, were just kids, and scared kids at that, anyone would be. From their earlier briefings his group had been instructed to assemble at a particular location. According to the intelligence reports there was an apple orchard, a good-sized apple orchard, just outside of Vierville a little distance away in an area named Gruchy. During the landing operations if they got separated, lost their crew, or lost their commander, they were supposed to head toward this orchard in Gruchy since it was their first designated assembly area. Each man had his map on him that showed how to get to the orchard. That was their first destination - from the time they left the LST they were on mission to reach that apple orchard at all costs. They eventually reached the orchard and were able to take a short breather so they would be able to continue toward their first combat mission several miles to the south. When they took a count, they discovered that they had somehow not lost a man in their own regimental unit.

On their way to the orchard along the cliffs above Omaha Beach they looked down on the beach and saw the hundreds, the thousands, of men whose lives had been lost during the past two days. Their lives had been lost by enemy fire or drowning as the weight of the soldiers' equipment pulled many underwater during their attempt to get over the side of the Higgins Boats to avoid being shot trying to offload. The dead lay where they were killed and were left down on the shore, proof that Omaha was still not a secured beach or else these soldiers would have been retrieved in preparation for burial. The image of their sacrifice had been burned into Birch's memory as he walked among the dead on the beach. As the day wore on, his unit's first "victory" consisted of reaching the apple orchard with no lives lost. It occurred to him that they were all frightened young men meeting in someone's apple orchard with night coming on, already spooked by what they had just seen and heard. Still, they needed to re-group, get their nerves under control, and set out for their first mission later that night. "Trust my training," Birch thought, "Trust myself."

They left Gruchy early in the dark early morning hours of June 8 marching in two columns with one on either side of the road. About 54 ki-lometers (33.5 miles) southwest of the apple orchard was their first combat mission, a small town named Insigny. They had not heard of Insigny, but it was now their primary objective. Plans changed. Instead of following the original plan of moving south in support of the general advance off Oma-ha, they were under new orders to capture this town. The 1st Battalion, which included his unit, led the way with the second and third battalions following, forming a column on either side of the road two miles long. Things seemed much more difficult and tense at night because it was im-possible to keep three battalions marching along the roadways quiet, and even more difficult to detect where their German counterparts might be positioned. They, and the tanks accompanying them, traveled through the night stopping once for rest, and resuming their advance around 0400 on June 8. They crept along and stopped whenever they spotted enemy tanks in the dim light.

This march at night, with the enemy nearby, hearkened back to a night years earlier when Birch was fourteen years old, and his family lived in Kentucky. A local elderly man was ill and not expected to live very much longer. Neighbors and family in the immediate area took turns sitting with the old man to keep watch and notify others if he worsened or happened to pass away. The night came when he and his dad, Levi, took their turn to sit with the man through the quiet hours of the night while most others were sleeping. In the early dark hours of the morning the old man passed away. Levi said, "Forest, take a lantern and go to Howard's house and let them know that Mr. Crabtree has passed." This meant making a two-mile trek in the early hours of the morning long before dawn. The woods and road that he was so familiar with during the day took on a different look and feel at night. Young Forest Birch clung tightly to his lantern and ran with haste as each shadow seemed to come alive, revealing motions not evident during the day, yet prevalent in the dark early-morning hours. He was much more aware of the cacophony of sounds, some quite loud, that occurred in the woods at night.

Now, years later, here he was, a man marching with hundreds of others in the middle of the night. No longer were mere shadows and sounds frightening him. Instead, each shadow was a potential hiding place for a real threat who not only bore weapons but had the intent to do him, and his fellow soldiers, harm. Each sound needed to be analyzed: natural or unnatural? Now, in the middle of the night, his life really was in danger, but he could not tell exactly where the danger lay. It was a new kind of fright this time that caused him to cling to his personal firearm rather than a lantern. It made him experience again the emotions he felt as a spooked teenager. The fear this time proved to be justified as the Germans attacked the column first while it was still dark and then again later after daylight. The fighting was merciless as the Germans fired on the Americans with personal firearms, machine guns and artillery guns called 88s (which were used in tunnels as well as mounted on tanks).

To make matters worse, about halfway along the road to Insigny the 1st Battalion stopped outside of a small town to wait for the 2nd and 3rd battalions to come together. The sun was up by that time, and it was easier to identify potential trouble spots. As the two battalions approached the town, they were seen by nine British fighter bombers who mistook them for the enemy. The bombers proceeded to fire on the column without mercy leaving six members of the 175th dead and eighteen wounded, not to mention the destruction of vehicles and equipment.[3] Birch was bitterly reminded of foul ups and mistakes that so often end up costing soldiers' lives. He thought again of the disaster at Slapton Sands. It was now almost too bitter to take, knowing this attack by the bombers and the resulting loss of life was completely in error and caused by the Allied forces. It was enough to do to fight the Germans, without having to take extra measures and precautions to ward off danger that arose by way of mistakes, miscommunications, and ineptitude. Worse still was the original order not to return fire on the Allied bombers, a decision later regretted, and that caused some commanding officers to order, "Fire on any aircraft, friendly or unfriendly, that fires on you first! We cannot conduct this war by walking on eggshells trying to worry about shooting back at Allied aircraft when they do not make the effort to assure their target is the enemy!" Yet on his company's daily report it only read, "Left Vierville bivouac area 0200 hours en route Insigny. Determined resistance encountered."

With their progress slowed by enemy presence and friendly fire, the regiment took the entire day on June 8 to approach near to Insigny. Early in the morning hours of June 9 they were at the outskirts of the town. Throughout that day they attacked and were counterattacked, but were able to fend off the German counterpunches, which proved to be ineffective. Eventually the American tanks moved in to disable the German artillery, and with the buildings housing them in flames, the Germans retreated. While the commanding officer of the 175th Regiment would have preferred a more cautious approach, the commanding general of the 29th Infantry Division decided on an aggressive approach to Insigny. This would set the tone for other missions of the 175th to come.[4] They continued to have their good days and bad days. Still vivid in Birch's mind was the beach. If there ever was what people called "hell on earth" that day was. He realized it was something that he would never be able to forget, not for the rest of his life. He certainly would not want to see anyone ever have to experience that. The sky was dark with planes and the barrage balloons, so much so that he did not see how a plane could fly there above the invasion force without having a collision. They were falling just as fast as they were being hit, ships were being hit, and it was simply unbelievable to him what transpired back there merely a couple of days ago. For the next ten days they fought bitterly in a whirlwind-like effort that required what felt to him like constant killing over and over.

With each ferocious battle and the associated loss of life, Omaha seemed farther and farther away, both in time and space. If there was any respite in war, he saw no sign of it. It required an entire day for his unit just to move a little distance, resulting in the unit moving only a few miles further inland as the middle of June fast approached. By then they were getting low enough on troops they already started to realize that they would very soon need to call up for reinforcements. He had become very well acquainted with the men he served with routinely, especially those of his own unit, but more and more he realized he was going to have to continually reacquaint himself with replacement troops. On the other hand, they would not be able to get reinforcement troops in place, let alone trained, until they could get the beach established so that it was completely safe to land.

The aggressive approach of the commanding general of the 29th Division paid off as far as results go. The price the division and its regiments paid in human life was higher though. Even in war time there is an economics. If you want to achieve your objective as quickly and as powerfully as possible, it will cost you more lives lost than if you pick away at it little by little. At least that was the way the economics presented when officers disagreed on the best approach, as was the case early on with the 29th Division's commanding general and the commanding officer of the 175th Regiment. The economics of war do not follow precise rules like the economics of corporations.

Hidden beneath the varying approaches taken in war were the unknowns that can alter outcomes instantaneously. One of these is the enemy's perception of you when you strike quickly and powerfully versus gradually and cautiously. A major concern was that caution on the part of the American forces could give the Germans the impression that the Americans were overly cautious and fearful of engagement. Such an impression might result in more aggression, more risk-taking on the part of the German military and in turn more loss of American lives. In other words, paying more up front by an aggressive approach had the potential to save greater numbers of lives later. When it came to the enemy they were facing in Europe, the division's commanding general was convinced the aggressive approach was the best, most effective approach, and that in the long run it would result in less lives being lost.

4. The Foxhole that Birch Built

FOLLOWING THEIR VICTORY at Insigny, the 175th Regiment started to settle into a front-line routine. Regimental battalions formed the front with the 1st Battalion assuming the forward most position at the base of this long front formed by the 1st and 3rd Battalions. As the front line of the war in France, the 175th continued its march south toward the city of St. Lo., located a little over 40 kilometers (25 miles) due south of Omaha Beach. St. Lo was a city with multiple intersecting roads, making it a valuable jumping off spot for making further progress into France, and eventually for advancing toward Paris. In other words, it was regarded by the Allied forces as a major strategic crossroad. The Germans also recognized the importance and value of St. Lo, and accordingly they made efforts ahead of time to reinforce defenses there, with the aim to keep the Allies from capturing it. In this way they hoped to trap Allied forces in the northeastern portion of France, thus halting their progress. All the soldiers on the ground knew for sure at the time was that they had another objective by the name of St. Lo, and like all the other objectives, it was essential to the success of the Allied invasion for them to complete their mission without fail. They continued onward with the aim to get to St. Lo as quickly and as safely as they possibly could.

Making forward progress, though, was very difficult for more than the obvious reason that the enemy was waiting along the way to slow them down. One of the most unique terrain characteristics found throughout Northern France was something known as the *Norman Bocage*, or the "hedgerows." These land features were first introduced by the Romans in ancient times. They were originally constructed as trenches with raised earthen sides. These were designed to be used for providing water for the crops and animals within the fields they surrounded. Throughout the centuries the earthen mounds lining the ditches became overgrown with thick, tall growths of various types of vegetation, thereby converting the ditches into tunnels with walls of vegetation reaching upwards of around sixteen feet above the top of the mounds. These walls of vegetation grew to become extremely thick and essentially impenetrable. They were nearly impossible to access, let alone pass through, at least not without heavy-duty equipment specifically designed for cutting a passage. The *bocage* was found nearly everywhere throughout Normandy, with the dense hedgerows forming what were essentially fort-like enclosures around the myriad fields. Though he was raised in the country and was familiar with farm life, Birch had never seen or experienced anything quite like the hedgerows. None of the Americans had. He tried imagining what it might be like to maneuver his way through sixteen feet high brier patches mixed with hedge apple trees, honeysuckle, thorny pyracantha bushes and a few other types of wild growth, all about as thick as it was tall. This seemed to be a rough approximation of what he and the others had encountered in Northern France. The *bocage* was not something he would be able to describe easily in a letter to his family back home; one really needed to see it in person. It simply sounded too fantastic to believe that it would be nearly impossible to maneuver through the thick undergrowth covering the mounds and trenches.

Not only were the hedgerows impervious to passage by individual soldiers, but they became barriers for vehicles, including tanks and other equipment needed for moving about on the fields of battle. Massive tanks could not merely run over or through hedgerows as they could other types and formations of vegetation. The hedgerow walls held soldiers in, kept tankers from moving in any direction they chose, and essentially trapped men, vehicles, and equipment inside the fields they enclosed. In this way these natural barriers forced anyone trying to maneuver them to travel the natural tunnel pathways. Unless they could find a way through the hedgerows, the Americans' movements in the *bocage* could always be anticipated. For the 175th and others worming their way through the countryside of Northern France, this is where the danger lay. These vegetation forts afforded the Germans opportunities to make advanced preparations for defending the fields in Normandy against the Allies, whenever they might attempt an invasion. To this end the Germans practiced maneuvering through the tunnels and chose the best locations for placing machine guns as well as anti-tank gun emplacements. Snipers practiced firing from trees into nearby enclosed fields. Most importantly for the artillery, the Germans planted stakes near what they believed would be the most probable routes Allied troops would have to take when trying to maneuver their way through the area. Then they mapped the locations of these stakes using a detailed coordinate system, thereby giving defenders the ability to call for artillery fire quickly by referencing the coordinates where the artillery barrage was needed. The result was that incoming fire could accurately hit pre-targeted positions occupied by the Allied troops. Before ever an Allied footprint was pressed into the sands of Omaha, there existed a well-mapped grid of the *Norman Bocage* that was available to be used for accurate targeting of artillery barrages against them.

Compounding the problem was the fact that the enclosed fields were of irregular shapes and sizes, rather than forming a proper, regular rectangular grid of parcels. This made it even more difficult for anyone trying to traverse them. While many of the fields were roughly rectangular, they came in a wide variety of sizes even among adjoining plots. All these fields were primarily connected by thin wagon trails winding throughout and among the irregular enclosures.[1] The key thing was that unless the Allies could find a way through the hedgerows, their routes of access were predetermined by the tunnels of vegetation, and every predetermined route was marked for immediate targeting. Solving this problem had to be a priority or they might be trapped inland much like the troops were trapped on Omaha Beach during the invasion.

These factors combined to render the situation perfect for defense, while at the same time making it a virtually impossible situation for invasion. In the first days of what became known as the "Battle for the Hedgerows" U.S. troops tried to attack the enclosures with full force.

Their aggressive strategy was to use speed and aggressive action to overwhelm the defenders. German machine guns that were already pointing directly at the openings used by the U.S. troops cut down the Americans instead. The Allies' advance was greatly delayed by the German hedgerow defenses, but the Americans eventually found an answer to the hedgerow problem.

A U.S. Army sergeant from New Jersey figured out a solution to the problem. He found that by welding a bar across the front of a tank and adding metal prongs across the bar, the tank would act like a plow and cutter. Doing this to a fleet of U.S. tanks allowed the tanks to push down and cut through the hedgerows and destroy obstacles without simultaneously exposing the tanks' weak points. These were the conditions under which Birch found himself, along with other members of the 175th, as they tried to make forward progress from one objective to another. He still could not get over how strong and impenetrable the hedgerows proved to be, how they had become deathtraps rather than access roads. He knew he would likely not see anything quite like it again, that is if he lived through this aspect of his tour of duty.

On June 13th the commanding general of the 29th Infantry Division issued an order in the hope of assuring that the 29th did not become complacent, as if it were even possible to do so considering how things had been going for the men of the 175th Regiment. The order was issued for every front-line rifle company, whether it was in a quiet sector or not, to send a patrol behind enemy lines every day. Furthermore, every battalion of the 29th was under orders to capture at least one prisoner every day. Many in the 1st Battalion were heard to moan and gripe out loud when they heard the order read. They deemed it too dangerous to try to take prisoners from behind enemy lines daily. The consensus among the troops was that the effort to capture German prisoners daily, cost too many lives for the little gain it achieved for the Allies. Still, the order stood and was one with which they had to live for the time being. Making what progress the unit could, given the conditions of travel, the men of the 1st Battalion arrived in the area near a village named Les Mesnil-Rouxelin on June 17. This village was located around 3.25 kilometers (2 miles) north of the city of St. Lo, which was their main objective. It felt like they were making some progress at last. Birch was on edge and yet relieved that he and the others had been successful thus far, even with the obstacles they had encountered. On their trek south, the 175th captured various pieces of ground north of St. Lo, and therefore experienced the most successes of any of the regiments in the 29th during the middle of June. More specifically, the 1st Battalion had broken though German lines on the way south toward St. Lo and had pushed more than a mile ahead of the closest 29th regiment. It looked to some as though the 1st Battalion was going to take St. Lo on their own, without any help. Now, just outside of Les Mesnil-Rouxelin, it was night, and their expectation was to move into St. Lo the following day.[2] The belief generally held by the Americans was that the Germans were on the run, and that though there would be fighting, the men of the 175th would have the upper hand.

Birch would never be able to forget the next day, June 18. The spot on the military map where they had stopped was designated as "Hill 108." Early on the morning of June 18 they could tell they were going to be in for a very difficult day, because the Germans began a relentless bombardment of Hill 108 using howitzer artillery along with mortar shells. It was then clear to the Americans what was in store. Following the artillery barrage a German infantry attack would soon follow. This was standard practice, and it also meant the Germans were not on the run. Soon the Americans were bogged down under heavy fighting by enemy troops, who were some of the crack, top-trained soldiers of the German Army. These German soldiers were able to hold the 1st Battalion of the 175th Regiment to a virtual stand-still. The entire unit absolutely could not make any forward progress. As the battle settled in, the men of the 1st Battalion found themselves lined up in place behind the fence that enclosed Hill 108, fighting during the day, and waiting for more incoming mortar shell attacks as darkness began to fall on them. Birch was not alone in feeling glum and heavy about the turn of events. There was no mishap, but they certainly had a faulty impression about the status of the German troops, who seemed more than willing to stand and fight to the last man, rather than behaving like soldiers on the run. The heaviness of heart affected the entire battalion which was now trapped and surrounded on Hill 108. Moreover, they were not merely trapped, they were beginning to suffer large numbers of casualties.

Thus far in their experience since landing on Omaha Beach, they had not seen anything remotely approaching the ferocity with which the Germans were now bombing and fighting them. The fear at night came like a dense heavy cloud hovering over them and making every breath labored and burdensome. This fear was mainly due to the nature of mortar shell attacks. Mortar shells by nature could not be heard, at least not during the entire time of their travel along their trajectories. Thinking about their dilemma while trapped in a field on Hill 108 Birch thought, "If only I could hear them, I might have some way to judge where they will land. But this soft popping sound over on the other side, and then nothing until it explodes over here is torture." He recalled their sergeant telling them the best thing to do is to protect yourself as securely as possible. Birch knew if you give into worrying about where a mortal shell might land, it can drive you nuts. The best advice was to follow your training, trust your instincts, and do the best job you can to protect yourself with the materials at hand. There is no value in worrying about those things beyond your control.

Mortars worked by being launched into a free fall trajectory, much like the way a baseball is thrown. The shell is lofted up and forward into the air and then follows the natural path of travel based on the angle and force at which it was launched. There was no way one could hear mortar shells to determine where they were going to land. This was very unlike artillery shells that could be heard from the time they left the barrel of the gun to the time they landed and exploded. So long as a soldier could hear an artillery shell, he could tell whether it was going to land close or travel further on overhead. Birch had learned to judge artillery shells to a point and got somewhat used to it, so that when he heard artillery fire, he became good at judging whether it was going to hit somewhere in the immediate area. Of course, one always took cover when artillery shelling occurred. The problem with mortars was that the shell made a soft popping sound when fired but then there was silence up until the exploding sound and concussion when it finally landed some distance away. It was the silent trip from "there to here" that made mortars so unnerving. "Those are the ones that get you" Birch thought. Those were the ones of which he was most fearful, and he was not alone. Everyone dreaded the soft popping sound of a mortar and then hearing nothing until it exploded on impact. All through the night the Americans and Germans played cat and mouse with only an earthen fence between them.

Another sound Birch dreaded to hear was the sound of the German Tiger tank when it fired its turret gun, which shot an 88 mm artillery shell. The Germans placed large artillery guns, called 88s on their Tiger tanks, as well as in permanent emplacements. They had one of the most frightful sounds that he had ever heard, so much so that he felt as though his heart nearly failed him upon hearing them fire. He once bemoaned to a buddy, Sergeant Long, that the Tiger 88 "…is a screaming shell that will absolutely put you out of your mind." Furthermore, the Germans could fire directly at you with them, shooting them straight like a bullet, or angle the shell to give it the effect of lobbing over to its target. Sergeant Long listened as Birch continued to comment on these strange and frightening weapons. Both men knew all they could do was try not to let it unnerve them or they might start failing at their jobs. Birch and his crew were going to be facing tanks, even Tiger tanks, and it behooved him to try not to let the sound of the Tiger cause him to lose concentration. He knew that doing so would most certainly save his life someday. Of course, all of this depended on their distance, but the German Tiger 88 was one of the most vicious shells that could be fired at troops. Birch still felt as though the sound of the Tiger firing its 88 would almost take one's sanity, and the Germans used those things frequently on the American troops. As he continued to think about the Tiger Tank Birch thought, "When that shell is coming at you, you cannot determine anything about it. Whether or not they shoot at you directly, you can never tell where it is going. The problem with the Tiger tank is that the screaming made by the 88 sounds like it is coming directly at you every time. Those things really make you need a psychiatrist." Sergeant Long and Birch talked about how they sometimes thought the Germans intentionally designed this gun to scare the life out of their enemies. If they could do that, then they might get an enemy soldier to take his mind off the task at hand. The trick was not to allow the gun "get into your mind," otherwise the Germans accomplished just what they wanted.

Clearly some weapons caused more dread than did others, which made it doubly difficult, and then some, to operate in the field without thinking too much about it. That was part of the design behind the Tiger Tanks' 88s, to make you dwell too much on it. The explosive shell was bad enough, but if the Germans could strike fear into their enemies when using this weapon then, whether it hit its target or not, it might cause the other side to make unnecessary mistakes. This is where a man's training would really come into play and prove its value. The training the Americans endured upon entering the armed services was designed to enable them to act quickly, automatically, without giving too much thought to a dangerous situation, which could cause them to begin to make mistakes. It is a good approach, but soldiers are just people who have everyday lives with aspirations and dreams before they become soldiers. Trying not to think too much about the dangers surrounding them on the battlefield was a constant internal struggle.

From June 17 - 19 Birch and his unit remained pinned down on Hill 108, an area that encompassed a very large field. Moreover, their battalion, which was normally made up of around 600 soldiers, was now down to a little over 300 remaining. The fences there were earthen fences made up of rock and dirt, solidly packed and sown in grass. To the Americans these looked similar to dirt mounds except that they were about six or seven feet high, and the soldiers could not see over them, not without climbing on them. It was rare to find one of these that you could look over just by standing on the other side. There Birch was, hemmed in by a fence too high to see over with the standard gateway built into it that neither he nor his peers could use. At night when he would try to bed down for sleep and quit fighting, both German and American sides continued to lob shells at one another or, if they were close enough, they would roll grenades over the top of the fence to each other. When the fighting between them died down a bit at night, the Germans would try taunting the Americans in the hope of causing them to respond in kind and reveal their positions. Though he did not understand German, Birch could hear the enemy talking to each other on the other side. Of course, it donned on him that they could also hear him and his fellow Americans talking on this side. They were nearly in each other's face. This went on all through the night and made the nights as tortuous as the days. It was nearly impossible to get any rest, which was so essential for being able to continue fighting the following day. The Germans were on one side of the fence and members of the 1st Battalion were on the other side. That's about how close they were to each other for what seemed like was going to be a very long exchange, with the Germans slowly increasing their advantage. Just so long as the U.S. troops were surrounded and trapped, the Germans could keep up the fighting and wear the Americans down until they ran out of supplies of ammunition, food, water, and eventually men.

With no way out for the 1st Battalion and no way in for the Germans, the men of the 1st Battalion bedded down and dug foxholes for their stay on Hill 108. These were to be designed both for a measure of comfort but mostly for protection. The men typically buddied up such that one guy would sleep, while the other kept watch. Their major was wounded, and they had no way to get him out. They could not call out using their radio because every time they attempted to use it, the other side would target them and resume bombardment of the Americans. The Germans also cut the Americans' telephone lines, so that they could not communicate to the back areas to relate their status to the other units during the battle. At night the men of the 1st Battalion stayed put, and soon learned the tiny little shovels they originally hated so much were dandy little tools one could use for getting some level of comfort in the field. These shovels proved to be essential personal equipment for all the soldiers.

As the battalion tried to settle in on Hill 108 Birch worked to complete his defensive "fighting position" otherwise known as a "ranger grave" or "foxhole." In an area that was partially forested this required considerably more effort on his part than it would if he had been doing this out on open land. Since Hill 108 was a large field, but also had many trees throughout, he dug his foxhole as though he were in the forest. First was that he had to concern himself with digging down then cutting and removing tree roots. He was glad that this was June because trying to do this on frozen ground would make it exceptionally difficult. While digging his foxhole he mounded the dirt around its perimeter. Then, as time allowed, he worked to gather tree branches, especially pine branches as they were handy, for use as cover for the hole. His preference was to cut down trees in lengths to span the dirt mounds on either side of the foxhole. Then he piled the branches and leaves over top of the logs and especially in the spaces between the logs. This helped to prevent dirt from falling or being blown between the spaces. To complete the top, he worked to place boughs over the logs in a dome shape, as much as possible, to help drain water away from the hole. Lastly, he tried to insulate himself from wetness in the ground by placing more boughs and leaves in the bottom of his foxhole. He had worked long and hard to be as comfortable as he could possibly be, given the conditions under which he was living.

As he worked to get things just right a fellow soldier came over to his foxhole and said, "Birch trade with me, I can't sleep over there." Trying to throw this off as nothing he replied, "Oh you'll be alright." As his height was mostly in his legs Birch worked with diligence to get his foxhole dug out to fit his needs. He wanted it to be long enough for him and worked to get it in pretty good shape. He covered most of the hole with some old timbers and dirt and had a place fixed where he could slide down into it with just enough space to slide in and out. He was finally looking forward to being able to rest, where he could lay down inside and stretch out his legs. He dug the hole long enough. He even dug a little hole to serve as a shelf inside, so that if he had a letter or something else from home, he could read it and lay it up on the shelf. As he neared finishing, he thought, "Nice home."

Still the other soldier would not leave it alone, but instead kept going back over where Birch was trying to finish up and pestering him about switching. Again, he said, "Birch, trade holes with me, 'cause I can't sleep." This was getting beyond frustrating and bordered on causing a fight, as Birch was dog tired and had put great effort into fixing his space. He did not want or need some goof off trying to make a scene and trade him foxholes just because his was better. He just figured the other soldier did not have as much coverage over his foxhole as Birch did over his. He had put the effort into it to protect as best as he could against mortar shells. He kept putting the guy off because in his mind it was just a scheme to get a better spot. Eventually however, out of frustration and to be able to get some rest, he finally replied, "Well I'll trade with you tonight. I know you can sleep in that spot as good as you can this one." He determined that this would never happen again, even if he had to get physical to make his point. However, on this night he was too tired and did not really like it when U.S. soldiers fought among themselves. It should be avoided, but sometimes you had to stand your ground and he was determined never to give in to such a ridiculous request again. This was going to be his first and last time to switch, no matter what the guy said or did later. Besides, they were only about ten or twelve feet apart as it was, so it did not make sense to him that the other soldier would be so persistent about a switch based on comfort.

He noticed something about the other soldier's demeanor. He did not really seem to be belligerent as much as he seemed to be nervous, restless, even ill at ease. This was also part of the reason why Birch made the one-time exception. Finally settled in, he and the men in his unit were resting as well as one could rest in a hole dug into the ground. In the middle of the night, long before morning light dawned, the Germans launched a dreaded mortar attack. During the commotion, as everyone tried to protect themselves as best as possible, an enemy mortar shell landed squarely in the foxhole Birch had built for himself and where the soldier who traded with him was now reclining. The man had no chance at all surviving it and may not have even known what hit him. No one could sleep from that point on, and Birch again experienced the old familiar feeling of weakness in his legs and clammy perspiration all over, like he felt when he saw the truck loaded with men disappear right before his eyes on Omaha Beach while he waited on the LST. This night as it swept over him it gave him a chill. Suddenly he felt very cold yet very much alive.

The battalion could not remain penned down in the field or else the Germans would wear down the entire battalion to the last man. One member of the 1st Battalion decided to break through, and travel back behind the line to get help. He did not personally know the man well, but from what Birch did know of him, the soldier had worked as a hunting guide back home in Maine. He led hunters who traveled up to Maine on game hunting trips and backpacked them out into the country. He was used to tracking animals in wilderness territory, and heavily wooded areas were a second home to him. Since they were pinned in, and were unable to communicate outside of the enclosure, this soldier made his way over to the major and did not ask but stated, "I'm going for help." The major replied, "I can't ask you to do that…I can't stop you, but I can't give the okay. If you go you go on your own." To this the soldier responded, "I'm going," and with that he slipped out without the major's permission, but also without any effort on the major's part to issue orders for him to stand down. The major, breathing deeply due to his wounds rolled his eyes to the medic attending to him. Both knew this was virtually a suicide mission. Silently the officer and medic each hoped the soldier would make it out to get help, and the officer knew he risked receiving all blame for not ordering the soldier to stand down. Neither said anything about it, for at this point they became spectators.

The first thing that came to mind about the hunting guide was that he did not seem to be an imposing figure. He was of short stature, red-headed and wore a mustache. The guide slipped away, and as they all later learned was able to make his way back behind the lines to get help. He was stealthy, as though he were one with the land. Though surrounded, he took his razor-sharp knife, put it in his mouth, and took off in the middle of the night. Later, early in the morning, they brought in fresh troops. Secretly on June 19 they woke up Birch and the others who were still there and told them to "Get up, be quiet, be easy rising up, stay quiet, make as little noise as possible, and ease on out."

The first streaks of light next morning revealed fresh U.S. troops awaiting the Germans. This time the Germans thought the Americans were nearly beat, only to find out they were prepared to fight. Overnight the 3rd Battalion replaced the worn-down, weary 1st Battalion. They held on through the entire bloody, bitter fight on Hill 108. When daylight came the men of the 1st Battalion discovered four dead German soldiers, lying where they were killed. The hunting guide stealthily crept up, unseen, unheard, slit their throats, and moved on. It seemed a cold-hearted, almost business-like approach, but that was the skill he could use to secure help. Later he was decorated for his mission. The resolve of the men of the 1st Battalion, with the skill of this soldier who had worked as a hunting guide, enabled the 1st Battalion and the rest of the 175th to go on toward St. Lo. It was crucial for them to achieve a breakthrough, so essential in the Battle of the Hedgerows. This fight on Hill 108 cost the 1st Battalion over forty percent of its pre-battle strength. Due to the high cost of lives, Hill 108 became known instead as "Purple Heart Hill." For their ability to hold their position on Purple Heart Hill for the entire day on June 18, the 1st Battalion received a Presidential Unit Citation as well as the French Croix de Guerre.[3]

5. Tar Heel Down

THREE STRAIGHT DAYS of very intense fighting had the members of the 175th, as well other units of the 29th, exhausted. The day the 1st Battalion left Purple Heart Hill it began to rain and rain hard. One of the battalions of the 175th noted in its journal that the weather was miserable, that all positions were in thick mud, and that it was the worst weather they had yet encountered since starting their campaign back on Omaha. As a result, June 19 became a day off from the fighting that had worn them down. Unknown to the men fighting there in the middle of Normandy was that the severe weather was much more of a problem than merely making things muddy and miserable. Meteorologists noted that the conditions in the English Channel at this time were the worst since the turn of the twentieth century. The sea rolled and churned for three days and the artificial harbor (referred to as a "Mulberry" harbor) was broken into pieces by the force of turbulent eight-foot-high waves. Also was the fact that strong winds wreaked havoc on the LSTs, and left hundreds of troops stranded back on the beaches.[1]

As far as Birch and his crew knew they had a day off from fighting the enemy, and spent it fighting the weather instead, trying to stay dry and comfortable as much as one can do outside in weather like they were experiencing. In the meantime, Allied armored tank divisions took out a fortified German position near Purple Heart Hill where the 1st Battalion had held their position at such great cost. As the tankers endeavored to move into St. Lo, they were met with the precision bombardment by the Germans using their pre-arranged coordinate mapping system they had developed prior to the Allied Normandy landings. This coordinate map marked all the strategic locations in the *bocage* so the Germans could accurately target Allied forces. The Allied tankers' efforts came to an end on June 30, when they finally had to be withdrawn and replaced by the infantry to complete the fighting to take St. Lo. As a result of these developments, every infantry battalion of the 29th Division was able to obtain a few days of rest sometime during the interval between the last of June through the first week in July. After what he and his buddies had just experienced on their way to St. Lo and then on Purple Heart Hill, Birch finally felt for the first time that he could truly stop and collect himself. "Maybe" he thought, "I can actually get a little rest."

While their experience on Purple Heart Hill was the worst that he and his fellow infantrymen had experienced since landing on Omaha, they were to face resistance every bit as bad near the end of July. July 30 turned out to be bad for the 175th Regiment and even worse for other units of the 29th Division. As the sun rose that day, the commander of the 175th had earlier set up the command center for the regiment at a crossroads village named Villebaudon, about 22.1 kilometers (13.7 miles) due south of St. Lo. A general rule was that any village or city that was a crossroads location was a valued military objective. In Normandy this was especially the case because off-road travel across the countryside was made virtually impossible by the *bocage*. As noon approached on July 30, German observers spotted the 175th command post and used their predetermined coordinate mapping system of the *bocage* to call in artillery bombardment to make a precise shelling on the established post. Their aim was very precise as the shells directly hit two of the command post tents and killed all but two soldiers. As the artillery barrage ended, the Germans followed up with the expected infantry attack. It was a ferocious attack potentially endangering the positions of all the battalions of the 175th. The two surviving soldiers were instrumental in pulling together other men in this headquarters unit to maintain some semblance of order. In fact, were it not for them the command post would likely have been overrun and all the associated personnel captured.[2] The Germans pushed the Americans back and nearly penetrated the area where the regimental headquarters were located. For a short period of time Birch and the other men of the 175th found themselves surrounded again by attacking Germans. It was an all-too-familiar feeling and still fresh in the memory of the members of the 1st Battalion.

For another regiment of the 29th Infantry Division the evening of July 30 took a disastrous turn. On the evening of July 29, when the 175th Regiment had stopped near Villebaudon, the 115th Regiment passed it by as it made its way south toward the small town of Percy, just 5.9 kilometers (3.7 miles) south of Villebaudon. Early in the morning someone noted in the 115th's regimental journal that they were under fire constantly from high ground to the south of them. Unknown to the men of the 115th, German tanks were hidden in the *bocage*, waiting for them. As the tanks emerged the Americans were caught off guard. In just a matter of minutes the tanks took out the 115th tank destroyers that were incorporated to support the infantry regiment, and then turned their attention directly to the infantry. At one point during the conflict American observers were able to reconnect phone lines and call back behind the front line for artillery support. While the Germans prepared and mapped detailed coordinate systems for the entire *bocage*, the Americans made efforts to prevent the hedgerows from being more confusing to the troops than they already were.

The U.S. Army produced very detailed aerial photos of the entire area along with 1:25,000 scale maps of Normandy. These maps were scaled such that each millimeter on a map represented 25,000 millimeters on the ground. They were detailed enough to show the locations of all hedgerows and farm lanes, and they were very accurate in displaying terrain contours. American artillerymen used stereoscopic magnifiers to examine the maps for specified targets. The effectiveness of this system caused many to say the Germans feared the American howitzers more than any other weapon in the U.S. arsenal. They were counter weapons to the German artillery which used the previously mapped coordinate system prepared by the Germans. The American howitzers made good hits on the German tanks attacking the 115th and caused them to eventually call off the attack and retreat. Still, by the end of July 30 the 115th ended up with more than 260 casualties including 60 dead. It was a devastating hit for that regiment.

During this same evening some distance away, while they heard the distant report of gunfire and the screams of artillery shells, the men of Birch's anti-tank gun crew looked for a good spot to stop for the night. Little did they expect that the next day was to turn out to be especially difficult and personal for them. On the night of July 29, in the immediate area around Villebaudon, Birch and his crew settled down into a local barn intending to get some sleep. Most of the French farms and vineyards were equipped with very large winery barrels and nearly all these farms made apple cider. Every barn Birch visited had these "great big barrels" as he referred to them. He had tasted a lot of homemade apple cider through the years, but he never tasted anything quite like the cider made on the farms of northern France. He was firmly convinced in his opinion that these northern French farms had the best cider that ever was made. In Normandy apple harvesting, rather than wine, was a major part of the economy, and it produced three distinct types of cider: *Cidre, Calvados,* and *Pommeau* – all of which were aged in large oak barrels. The Americans took a liking to Normandy's apple cider, but they also had to be careful of other types which had greater potency to incapacitate a soldier due to the alcohol content.

It was such a barn that Birch and his crew used as their quarters for getting some sleep the previous night. The next morning, July 30, his gunner was shaving. In the field the U.S. soldier typically shaved using his helmet for a wash basin, and then usually placed a little mirror over the helmet on a shelf, or hung from a post, etc. His gunner was from North Carolina and as he shaved, he looked over to his group leader and made the following comment, "Birch I feel like my luck's about run out." Birch was well acquainted with such feelings. He recalled watching the men and truck disappear into thin air on Omaha, and the man with whom he switched foxholes blown to pieces in the middle of the night just north of St. Lo. Each time the familiar eerie feeling seemed to present itself and would have made it very easy for him to fall into the trap of judging that he would not make it, that his time had run out, that this would be the day he was destined to die. It was this very thing that his training had been designed to overcome. It was important to be able to function, almost like an automaton, and avoid falling into the trap of thinking too much about one's circumstances or what could happen, and then begin to fret about one's family, friends, and life prior to the war.

On the other hand, neither Birch nor any other soldier in the field was an automaton. At best his training could help him to remain focused so that by not thinking too much about it he would be less likely to make mistakes. He had been trying to rely on his training, but in times of quiet or at those times when he was settling in to try to get some sleep, his mind became very loud and active in thinking about it. He could not merely forget about what was happening and then move on. He had already seen and heard too much to dismiss it from his mind altogether. In addition, there was his life and family back home; when things were quiet, he could not help but think of them. There was probably not a man who had not, at some time or another, felt that old chill inside that came as a harbinger of doom. Birch wanted to acknowledge that he understood what his gunner was feeling and yet he wanted to help his gunner not to place too much stock in such feelings. Therefore, he replied, "Oh don't feel like that. I feel like that all the time." He tried to make his remark casual. At the same time, he wanted to express an understanding, or at the very least a recognition, of what the man was feeling and an acknowledgment that it was normal to feel this way at times. They chatted a bit as the gunner continued to shave.

He had no more than finished shaving when some army troops and infantry from another unit came rushing back through the area and their lieutenant came into the barn and asked, "Where are you guys from?" They told him and he said, "Hook that gun up and get out of here!" He was not their senior officer so nobody in their unit placed too much stock in what he said. Then, after a short while, somebody else came rushing through and said, "Get them guns out of here, the Germans are coming!" They told the crew that the Germans were pushing the Americans back from their current location. When Birch and his crew saw that they were serious and meant it, that this was no whim of an officer from another group, they jumped up and ran to the gun towed by the truck, loaded all the crew up on the truck and prepared to leave. On the truck everyone had a certain place to sit. All the men were to get on after which Birch and his gunner were to get on and sit on the ends in the back of the truck, one on either end of the two rows. This was so that they would be the first two men off, and the last two men on. Therefore, loaded up and ready to travel, they started back down the road in the direction from which they had come the night before.

On the way back down the road they came to an area where there was a crossroad, up ahead of them a short distance. Somehow the Germans had managed to get in behind the U.S. troops and set up machine guns at the crossroad, which were hidden to the men on the truck. As they started down the road and approached the location of the machine guns, suddenly bullets started flying, coming down toward where the truck was now positioned, and lighting up the sky with fiery tracer bullets. Every other bullet was a tracer bullet and when fired from a gun it produced the appearance of a streak of fire. Tracers gave a much more ominous look to the exchange of gunfire, and in a real sense put the "fire" in the term fire fight. From the perspective of the men in the truck the streaks of fire were coming directly toward them.

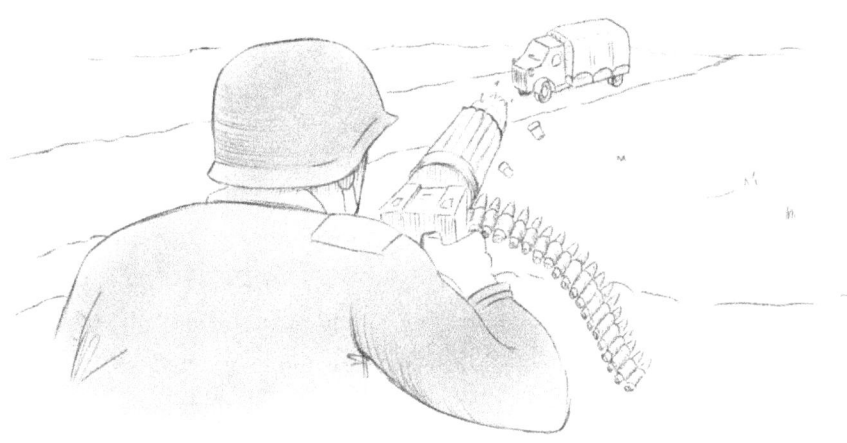

The driver suddenly slammed on his brakes, and everybody jumped out of the back of the truck. The men's training built into them the habit of jumping off and falling or rolling directly down into a ditch by the roadway, or to get as flat as possible on the ground. Birch was on the outside and saw a very large hole in the bank on his side. As trained, he jumped off the truck and headed for that hole. It looked to him like a safe place and as he hit it, he felt as though he was just made to fit right back in against the bank. However, for reasons he did not know, and would never know, his gunner did something completely against his training. He jumped off the back of the truck, but instead of rolling down into the ditch on his side as he was trained to do, he ran around the gun (a 57 mm anti-tank Howitzer towed behind the truck) on the outside, down to the driver's side, kept running on to the front of the truck, and then ran in front of the truck. All this time their sergeant was there yelling, "Get down man, get down man!" Then just as the gunner crossed in front of the truck, he took four bullets that Birch could count right across the middle of his chest. Birch heard the – thump, thump,

thump, thump – four dull sounds as the bullets impacted his gunner's chest. The gunner ran directly in front of the sergeant and into the line of fire of the German machine guns.

Their sergeant's last name was Fardie. He was from Boston, and in his Bostonian accent he hollered over to Pfc. Birch and said, "Birch, get over here," to which he replied, "I'm not coming over there." Fardie said, "We got to see whether Jimmy's dead or not." In a solemn tone Birch responded, "Well you can tell he's dead, he's not moving." Laying there in place Fardie again called, "Birch, get over here," with the same response, "I ain't coming over there!" As the machine gun fire began to subside Birch and the others could see soldiers, some of their own American troops, down near the area where the machine guns began to unload on them. Still, he was not about to make a move. In the meantime, over a period of just a few minutes, all the men of his unit continued to lay as flat as they could in that ditch, even though they could see their fellow soldiers and could hear men talking down at the crossroad. No one wanted to be the first to move out of the safety of the ditch. Then one of the men over across the road from him hollered over and said, "Birch you better get out of there!" to which he replied, "Why?" The soldier said, "Look up over your head." When he looked up, he noticed that some of the U.S. troops, and tanks had moved in and were cleaning out things down at the crossroad. One of the U.S. tanks had pulled up to the fence line along the road where the men from the truck lay and had his gun turret moving around directly over Birch's head. Had the tank fired that gun, the flash from its turret would likely have burned Birch to a crisp. The muzzle flash on the tank put out so much flame that hedges or undergrowth in front of the tank were likely to start burning after a shot or two. When he saw that, there was no hesitation, he sailed across the road, landed right in the ditch, and crawled down to Sergeant Fardie's location. Then together Fardie and Birch climbed out of the ditch and could see American soldiers standing out in the road down at the crossroad. Together they undertook the sad duty of going out to the middle of the road and checking on Jimmy, his gunner. No one could possibly know, nor would anyone ever know, why the gunner ran around the gun, down the side of the truck, and across the front of the truck in front of Sergeant Fardie. "What was he thinking about when all of this happened?" Birch wondered, as he said to Fardie "Jimmy had a feeling, a sense I guess you might say, that his time was up, that he would not make it through the conflict." "When did this come about?" asked Fardie. "Well, Sarge," replied Birch, "To tell you the truth he was just telling me this morning while we were shaving, immediately before they came through and yelled at us to clear out of there." In his mind Birch still wondered, "What could possibly have driven him to completely abandon his training and make himself the most visible target of anyone on that truck? Could it really be that thinking too much about the possibility of dying caused Jimmy to make such a mistake?"

The crew had to pull themselves together and continue with their tasks. Some of them just needed to distance themselves from the scene of Jimmy's death. They looked around and walked down to where they had to cross a depression in the road, and then on down into a deep gully in the area. Shortly afterwards they saw a jeep coming their way, and as it approached them, they saw that it appeared to contain someone important, clearly an officer's jeep. It turned out that riding in the jeep was General Patton with his driver. He rode in, pulled up to the men and stopped. He viewed the area around him, and then looked directly at Birch and his crew that were there and asked, "What outfit you guys from?" They told him and he replied, "Well what are you doing down in here?" Someone in the crew spoke up saying, "We just lost a man," to which Patton asked, "Lost a man?" One of the other men replied, "Yeah, he's laying out there in the road." In a cocky sort of tone Patton replied, "Well nobody was shooting when I came in." Naturally this sounded as though he did not believe the men, or even care what they had just told him. Birch had heard of Patton's manner and wondered whether what he heard was true or whether it had just been built up over time like a growing military fish story.

Now what he heard from Patton sounded like the vintage Patton he had heard so much about. It may indeed have been the case that Patton was not being shot at when he came into the area, because he came in a different way. He stood there by his jeep and chatted with the men a minute. Patton was known for his crude language, and this certainly proved to be the case as he talked with them for a while. Finally, Patton looked at the crew and said, "Well, anyway I hate it you lost a man." With that said he got back into his jeep and took off. General Patton was passing through this specific area because he was making the push for St. Lo, which was the first big city taken by the Allies. Then from St. Lo he and his "wild tank drivers" made their push through France. They moved so fast from that point on that the infantry could not keep up with them and did not even try to do so. The pace was just fast with no other gear. He was headed for Paris and so he was gone, regardless of who could or could not keep up with him. There were soldiers who marched where Patton had been and cleaned up various problem areas called "hot pockets." That was Birch's knowledge of, and experience with, General George Patton. He never expected to meet Patton and given the circumstances under which they met it was anticlimactic to say the least. He appreciated the expression of sorrow on the general's part though. That was something.

He and his crew continued back out to the road and retrieved everything off their gunner they were required to remove. All his personal belongings, along with one dog tag, came off him. They left one dog tag on him, put his bayonet on his rifle, stuck his rifle in the ground, and placed his helmet on it so the others would pick him up. They left then, as they had orders to move out. That is what they had to do. If they lost a man, or if they found a man, they removed his personal belongings, took one dog tag, stuck bayonet on his rifle, and stick it in the ground, the butt sticking up with the soldier's helmet on it. The Americans had crews who came along behind and took care of those left on the field of battle. These crews turned the information in to the commander, who then turned it in to headquarters, who notified the relatives. Quickly that became part of their routine, and one that Birch never got used to performing. Jimmy was back behind them now laying still on the field of battle, his death still a mystery given his strange behavior that day. In the time it took Birch to hear – thump, thump, thump, thump – Jimmy's life had ended. It seemed surreal that one minute he was sitting talking to Birch on the truck, and the next he was silenced by four bullets. Still, he would never be gone or forgotten in their hearts and minds.

Though Birch and the others knew that there were troops tasked with coming along afterwards to take care of the fallen, nobody really thought about what these "Graves Registration Service" (GRS) troops were really tasked with doing during the war. They were to care for the deceased soldiers, but they were also tasked with doing quite a lot for their living brothers-in-arms, and therefore for their country. For example, in the event both dog tags were missing, these troops would take prints of all ten fingers as well as prepare a dental chart. If the soldier's body happened to be in very bad shape, they were trained to make an extra effort to inject fluid into the fingers to obtain usable prints. In extreme cases they had even gone so far as to remove skin from the fingertips for prints. They also used other means for identification such as personal effects like various documents found in a wallet, as well as statements given by soldiers who knew the soldier who had been killed. Even things such as laundry marks on clothing could serve as crucial clues, as they listed the first letter of a man's surname and the last four digits of his service number.[3]

In addition to establishing identification, the GRS troops were responsible for making inventories of personal effects, such as rings, wallets, watches, photos, etc. These items were then shipped to the Quartermaster Depot in Kansas City, Missouri. There they were cleaned up and sent to the next of kin. In the field, the troops destroyed bloodstained items and anything else that could possibly embarrass or upset a soldier's family. Then they distributed perishable items like cigarettes, chewing gum, and rations to other troops. In addition, they also gathered government-issued items such as weapons and ammunition and distributed these to other soldiers in the field who had need of them. Great care was taken to honor the fallen soldiers and treat them with great respect and dignity, but also to help their friends and loved ones to avoid greater hurt during such a difficult time.

The GRS was also the team tasked with the responsibility for locating suitable sites for the establishment of cemeteries. They were relied upon for making an examination of the terrain of interest, including such detailed information as soil quality, as well as distance to enemy lines. Once they settled on the selection of a site, they had to lay out a cemetery grid, where all grave sites were plotted and mapped. They used tents for creating spaces where they could process the remains of the deceased. Typically, local civilians were brought in for digging graves and burying the dead. This was the process put into place by the United States Armed Services. So it was that the next unit of brothers-in-arms to come along to meet up with Jimmy, the gunner from the 1st Battalion of the 175th Regiment, were members of the GRS. These troops, like so many of their fellow soldiers who participated in the landings at Normandy, started off the war in Europe inundated with work and never able to catch up.

6. A Star in a Tunnel

In 1944, DURING the war, a new medal appeared in the military awards catalog. This new medal, the "Bronze Star," was an idea that originated with Colonel Russell Potter Reeder, a veteran of the First and Second World Wars. Reeder came up with the concept for this specific medal after he suffered an injury in Normandy which resulted in the loss of his leg, thus requiring that he be medically retired from the Army. He was personally awarded the Purple Heart, the Distinguished Service Cross (the first issued at Normandy during the Second World War), and Silver Star after his return home. Still, he desired field commanders to have a way to acknowledge troops in the field immediately, without having to wait until they made it home, assuming that they were able to make it home. The medal he had in mind was intended especially for those who had been in the line of duty for an extended period. Reeder initially referred to this new medal as the "Ground Medal." His idea became a proposal which traveled somewhat rapidly through channels until it eventually received the attention of President Franklin D. Roosevelt. Subsequently, in February of 1944, President Roosevelt signed it into an order as the Bronze Star Medal. The order in final form included a retroactive precedent for armed service members that dated back to America's entry into the Second World War after the attack on Pearl Harbor. Therefore, in 1947 the Bronze Star retroactively replaced both the Combat Medical Badge and Combat Infantryman Badge, both of which were still being used during the Second World War.[1]

Since its inception, the Bronze Star served to recognize those armed service members who demonstrated heroism in the field, or who were deemed to be meritorious in the performance of their duties. In addition, it became the fourth highest military award given to an individual. Qualifying for the medal involved performing acts during armed conflict against an enemy of the United States. Because they were already recognized by the Air Medal, meritorious or heroic acts performed during flight operations were not recognized using this new Bronze Star medal. There were three categories for which the Bronze Star could be awarded, which were valor, merit, and achievement. The Bronze Star medal itself was designed as a five-pointed star, 1.5 inches in diameter, shaped out of bronze and hanging from a red ribbon with a vertical blue stripe bracketed by thin vertical white stripes on either side. It was also a personalized medal in that its reverse side contained the recipient's name stamped into the middle of a circle surrounded by an inscribed phrase which read, "HEROIC OR MERITORIOUS ACHIEVEMENT." Like the colors of the American flag, the colors of the Bronze Star ribbon (called a drapery) represent the values of purity, perseverance, and bravery.[2]

The Bronze Star eventually came to have great significance for Birch and those with whom he served throughout the course of their tour of duty. He had heard of the medal, along with the many others awarded to soldiers for their performance in the field of action, and while he always deemed that it would be an honor to be recognized for your sacrifice by way of medals, he remained focused on doing the best job he possibly could do to survive the war so he could go back home. In his mind arriving home in one piece was the best award he could imagine receiving, better than all the medals in the military awards catalog combined. He felt strongly about this, especially given the high number of casualties his battalion and the entire 175th Regiment were experiencing. Still, he admired many of those who were highly decorated with bright colored ribbons and associated medals for their service.

It was now October 1944. He and his crew had been all over France, much further than they had expected. This was due in part to the fact that they were called to travel to the westernmost region of France, called Brittany, shortly after their encounter with Patton in the area around Villebaudon. After spending about a month engaged in heavy fighting in

Brittany, they were again called back to front line action, a routine they had already developed and maintained up through the latter part of August. This time though they rejoined the front line in Germany, leaving France behind.

On this bright October day Birch and his crew were relaxing on the side of a hill near Geilenkirchen, Germany, just 18 kilometers (11 miles) east of Sittard, Holland. They had parked their truck and gun along the road just off to the side and were waiting for military columns of men and vehicles to move forward. As they took this brief rest some of the men talked among themselves as Birch fell into thought about the developments of the past couple of months and the effect these were continuing to have on the 1st Battalion...

He thought to himself about the fact that he was now regarded as one of the older, experienced soldiers, even though he had just recently turned 22 years old. It occurred to him that here, now, in all this mess, something occurred to make him the one to whom others came for advice. He was learning an important lesson that experience, not merely time, was the measure of a man's age on the battlefield. The more experience one obtained during war indicated he had learned how to stay alive, at least to that point. An unwritten rule was that the one with the most experience was the one to whom others should listen and follow. He recalled that during their time in Normandy his unit almost constantly continued to fight long and hard, encountering resistance every bit as challenging and devastating to their battalion as they had back on Purple Heart Hill (Hill 108). As he approached the month of July of 1944 it occurred to him that it was his birthday month. Thinking about his birthday he recalled, "I spent my birthday on July 13 in a position east of St. Lo. We tried our best to take control of the city, in an effort that proved to be much more challenging than our U.S. command had anticipated. Why was that? Was our intelligence faulty, or were we just flat outmatched by the Germans?" On July 13 they had not moved too far from the position where he and the rest of the 1st Battalion were encamped north of St. Lo around the middle of June where they were eventually surrounded by the German infantry for three days. Though the men of the battalion held their position and some of them made it through, many of them did not. He thought about this and concluded that the losses they incurred during the battle for supremacy on Purple Heart Hill were so great it simply boggled his mind that any of the members of his unit made it out in one piece. In addition to the large number of fatalities, there was a great number of men who ended up wounded during this battle and were probably still battling the emotional toll that it took on them. Birch knew some of his buddies in Headquarters Company were still dealing with the trauma, but he thought it was probably best that he should not pry by asking them about it. Still, he wondered, "Who knows when it is best to say something versus when it is best to keep quiet?"

Though he was not one of the wounded, Birch struggled with the emotions that were dredged up when he thought back on his time on the infamous hill. In reflecting on it he knew that he had tried diligently to rely on his training to help him avoid falling into the deathtrap of introspection and worry. He was grateful that he had the strength to continue to fight and help his brothers in arms as their strength in numbers continued to dwindle. He thought, with a thankful heart, about the stealthy hunting guide in their unit who was able to go for help. "Wow, am I ever thankful for that guy and the skills he brought here to the war in Europe and especially to our unit."

As he and his unit continued to fight, toward the end of July the 175th encountered very strong resistance and for the second time the regiment logged hundreds of casualties. "This makes it more difficult because constant battles with large losses robs you of the time needed to think things through. I just wish I could sit with what I have been through for a while. I bet all the men do," he thought to himself. The daily unit reports for the 1st Battalion between the middle of July through the later part of August listed many who were either lightly wounded in action or killed in action. This time there were not as many dead as there were after the encounter on Purple Heart Hill, but this time he and his own gun crew lost a man of their own.

He could not help but think about Jimmy and wondered how his family had received the devastating news. In thinking about this he concluded that naturally Jimmy's family would probably like to be able to talk to some of his fellow soldiers who knew him and fought with him to ask if he suffered, whether the fatal blow was quick, and/or how he displayed his bravery through it all. Though he did not know Jimmy extremely well, Birch would have liked to try to answer his family's questions. After all he felt like they were getting better acquainted, and he was the one the gunner confided in the morning he was killed. Birch thought to himself, "Oh how I wish I could have said something to help him avoid the mistake of running directly into enemy fire." He needed to quit reminiscing about this and not fall into the "what if" pit that has no bottom. Talking to Jimmy's family was not possible; they would have to deal with the news on their own, and he hoped that the family was close-knit and strong enough to lean on one another in trying to get through it.

Birch was glad that he received his second promotion on August 8 from Private First Class (Pfc.) to Corporal (Cpl.). Around the time of his promotion, the 1st Battalion had made it past St. Lo and finally moved south of the city on its way to the next objective. On the very same day he was promoted to corporal, his battalion was located at La Moignarie, an area located about 38 kilometers (23.6 miles) south by southeast of St. Lo and near the small town of Vire. From that day through August 22, the 1st Battalion fought its way to St. Jean des Bois, about 24 kilometers (almost 15 miles) south of Vire. Again, the daily unit reports reflected many names of men wounded or killed in action. Birch thought to himself about the lonely and often perplexing fight in the *bocage* of Normandy. For those past weeks his regiment had been involved in the most severe fighting on the front line of war. Yet it seemed as though they had been forgotten by the world. He kept hearing bits of news of other, more visible battles being broadcast far and wide. His unit really was in the very heart of the activity in Normandy, and they probably touched nearly every major aspect of the war in France. Birch sometimes wondered if anybody even knew or cared about the 175th. He had to stop thinking too much along these lines or he might fall into a frame of mind unhealthy for a soldier on the field of battle.

He then thought about the recent losses. From August 8 through 10 the regiment lost around 250 men; he had not realized that at first. For some reason it seemed like they and other units of the regiment were always being placed in positions of near impossible odds that kept whittling down the battalion's overall strength. Then there were the turnovers; there was constant turnover, which made it so much more difficult to get to know the men, let alone effectively work with them. This could be a dangerous situation, and he knew they had several new replacements at that time in Headquarters Company. One had to get to know the replacement troops to a certain point or there would be no possibility of working together efficiently. It was not that Birch wanted to make friends. Rather it was a matter of life and death. One had to be able to anticipate his buddies in the field, trust them, and be certain of what they were capable of doing, as well as what they were not capable of doing. One also had to know what they were willing and unwilling to do. The turnovers made this more of a challenge, but it had to be dealt with if he and the others were to successfully complete their missions and make it through the war alive. He questioned on more than one occasion to himself, "I wonder how many casualties that our regiment will have suffered when the conflict is finally over, and how this number of casualties will compare to the size of the entire regiment as a whole?"

Up to this time no respite had come for his unit. Furthermore, on August 11 the 175th was assigned the task of pushing further forward to help one of their fellow regiments of the 29th. This regiment was caught out ahead of the rest of their division and became trapped in a fight with the enemy. This turn of events forced them into a situation where they were fighting on their own. The 175th worked to spearhead the attack to bring relief to their brothers in the 115th Regiment, and subsequently were ordered to continue further to the south where they met with very stiff resistance by the Germans. This was quickly becoming part of the continuing story of the battles across the entire Normandy front that re-sulted in such large numbers of killed and wounded in their regiment. The 175th learned this by experience repeatedly as their thrust to the south on

August 12 cost the regiment a total of 160 men in a single day. By the end of the same day the Germans were overpowered by the combined forces of the U.S. infantry and armored units. Even Patton's tanks were starting to roam around freely behind enemy lines.

On August 17 the men of 175th finally received news that they would get a real break from fighting and were subsequently pulled off the front lines for the first time since they had landed on Omaha. He never asked his superior officers, but the question came to Birch's mind, "Are we always being given these difficult assignments because we are the best ones able to accomplish them, or are these actually desperate moves the U.S. Army is making because our options are limited?" He thought about the long-awaited break they were promised, given, but ended up being very short lived. On August 22 the 175th was loaded onto trucks to travel to Brittany in the westernmost region of France 338 kilometers (210 miles) away.

In some ways this was an entirely new war zone for them. There were still hedgerows, but the soldiers were in an area of France with which they had to familiarize themselves anew. Birch recalled that as it was in Normandy, the German infantry in Brittany, even when surrounded, showed a high level of motivation when they fought. They did not seem to suffer from the demoralized, defeatist frame of mind some U.S. leaders expected early on. Again, the thought plagued him regarding whether this was another problem with U.S. intelligence or whether something changed the Germans. He thought about the enemy defenses, which were substantial. There was no room to bypass them there by the coast, as there was in Normandy. This left the 175th one plan of attack – a full on frontal assault. On August 26 the men of the 1st Battalion were again assigned a difficult task, this time to help with the capture of the area known on the maps as Hill 103. The resolve of the 1st Battalion that was tested severely on Purple Heart Hill (Hill 108) was tested again in the middle of June on Hill 103.

Birch was still amazed at the robust German defenses in Brittany which allowed them to withstand attacks of great force, but the Americans exhibited great tenacity. The result of this was a lot of hand-to-hand combat along with the exchanging of grenades. In this way the battle on Hill 103 was akin to what they had experienced earlier on Purple Heart Hill. The fighting was ferocious and the more the Americans tried different tactics, the more the Germans withstood them. Eventually, by the afternoon of August 30, the Germans were finally overcome with night attacks by the Americans who used the Bangalore torpedoes that were so effectively used on Omaha to burst through enemy fortifications and barriers. Even at that, the Germans attempted a counterattack later, but the Americans managed to keep Hill 103 in their own hands. This was the type of fighting Birch's battalion saw again at a location known as sugar-loaf hill, as well as other objectives of the 175th, until orders finally came down that the regiment was needed back at the front. Subsequently, on September 29 the 1st Battalion moved by train and motor vehicle to Holland, traveling 1086.5 kilometers (675 miles) to arrive in Sittard, Holland the next day.

Suddenly Birch was brought back to reality when Corcoran, his loader, kept calling his name. It had only been a few minutes but to Birch it seemed longer. They were still waiting for the columns to pass and had made sure to secure their truck and gun underneath some trees to prevent aircraft from spotting them. Earlier on in the conflict, one of their very own Allied airplanes shot their truck and blew it to pieces, obviously mistaking it for the enemy. The problem with this friendly fire was made worse by the fact that when it occurred Birch and his crew had laid out all the proper markers with the appropriate bright blue and orange colors so any Allied aircraft would be able to see that they were friendly. They watched as some Allied planes flew over but noticed one in the back of the pack who trailed off, turned, and came toward them, swooping down from his original altitude. They could hear the whine of the engine as the pilot accelerated during his dive. "That idiot's going to attack us!" one of Birch's crew exclaimed. They all were in disbelief that one of their own pilots would fail to identify the bright blue and orange markers. "Spread out and hit the ground!" hollered Birch. They dove for cover and the lone pilot fired his machine guns at their truck and anti-tank gun. They did everything right and still it went wrong, another mishap that resulted in the destruction of their truck and very nearly killed some of them. The fact that none of them were injured was the only good thing that came out of this horrible mistake. The quick thinking of the crew enabled them to save themselves while the airplane obliterated their truck right before their eyes. From that point forward they tried to hide their truck from enemies and friendlies alike. It made things doubly difficult because they had to watch for fire from anywhere and everywhere. It was always in the back of Birch's mind, and he even started reminding the men, "Watch out for the foul ups; they are not just frustrating or aggravating, they are deadly as well!"

On this day they were laying around on the hillside, enjoying the shade and talking, and still waiting for the last of the columns to pass. It had not really occurred to Birch just how much natural beauty served as the backdrop for this war that always seemed to squelch it. In fact, this was true of a great many wars. Some of the deadliest, most violent encounters have occurred in some of the most beautiful of settings. Reclining here on the hillside he could see the splendor of the foliage, with its shades of red, orange, yellow, and mixtures of all. Were it not for the machinery of war, with all its associated paraphernalia, he would have felt comfortable in this farming country with its small hamlets, country roads, and life lived off the land. It was true that the language was different and new to him, and the farms did not look like those back home in the U.S. Still the land, foliage, fields, and hard work that the people exerted to make a living were very much like what he was used to back home. Looking around and taking in all the beauty of Fall here in Germany he would very much have liked to take a hike across the countryside just to experience all that he could of one of Germany's most valuable natural treasures. It occurred to him that it would not take too much to make one forget about the bitter feelings that existed between the sides fighting against one another here. In many cases he could see the faces of his enemy that held this animosity, which he and his fellows in uniform reciprocated. As it turned out though, he rarely had the opportunity to look at the beauty of his surroundings. Rather he usually had to look past the beauty of the setting for the purpose

of assessing whether hidden dangers lay within and behind the beautiful
vegetation that seemed so welcoming at first glance.

As he thought about this, suddenly a call came for him to bring
his gun and loader up to another area. The sergeant on the other end of the
phone ordered him to load up the 57mm tank gun and bring it forward for
a special mission. Birch loaded up and left immediately and determined
that based on the location he would need about 30 minutes to get there. He
wondered why they wanted just one gun up there, because anti-tank crews
typically set up three guns in a triangular configuration to have the best
chance of disabling or destroying a German tank. One gun was essentially
useless unless you happen to make the perfect shot and hit the tank from
behind, which was a "one-in-a-million" chance. This probably meant that
they were not calling him up for the purpose of fighting a tank, nor would it
have anything to do with him personally, since they asked him to come up
and bring his gun, which meant he also needed his loader. He would just

have to drive over with his loader and find out for himself what he was being called on to do. That was the way it occurred at times, you received an order or a call, but with no clarifying information. To some degree he understood why. It simply took too much time to explain everything when the action was happening live every minute, and things were changing just as fast. Still, it would have been nice had they briefly told him what they needed, or how they thought he could help. To get where he was ordered to go, he and his loader had to travel through some relatively rough terrain, after which they had to maneuver truck and gun up onto a hill to arrive at the destination. Once there they saw a group of officers and men gathered and observing some sort of activity across the way. When he exited the truck, he could see the group was looking at American and German troops some distance away, across a long field.

The fields in Germany were different than northern France in that they were not enclosed by hedgerows. Here it was much easier to see the countryside at a glance, without feeling as though you were completely surrounded by walls of greenery. The reason for this was that there was no *bocage* in Germany, just more Germans. Birch stepped up to where they were looking across this long, gently rolling field and looking more closely he noticed that out in the middle there was a low mound formed by a natural rise in the terrain. On his side of the rise was a group of U.S. soldiers who seemed to be hunkering down, not moving forward or backward. They were essentially frozen in place. As it turned out, he was informed, the group was B Company and they were endeavoring to attack a German position on a hill across the way, on the opposite end of the field. As he looked, he could see that there seemed to be an opening of some sort in the hill; they could all see the opening. Most likely this was another tunnel built into the hillside and if it was like so many others the Germans had designed, it led to a bunker built into the hill that could accommodate various types of equipment and large numbers of soldiers. The bunker would have more tunnels running underneath the hill, forming an intersecting hub from which the Germans could operate quickly and with few repercussions. This situation was seen near the beaches of Normandy, over on Brittany, and several places here in

Germany. The German soldiers in the tunnel had the ability to come out of the protection of their tunnel, pin down B Company, and then retreat to their robust cover. As Birch and Corcoran, his loader, arrived, the Germans had the company pinned down out in the open field. The men of B Company could not come back, and they could not go forward, because of the little high place out there in the open.

This lone rise in the otherwise flat field was all that was saving them from being killed or wounded by enemy fire. This small mound was the only obstacle between his fellow soldiers of B Company (Birch served in Headquarters Company) and the fortified tunnel that was heavily armed and manned. The Germans could not get their artillery out far enough on a rail to kill the Americans. If B Company laid flat against the high ground, they were alright, but they could not make the move to come back or forward, or else the enemy fire would get them. This is the reason why those observing the situation called for Birch to bring the anti-tank gun and his loader. At the designated location where they had parked the truck there was a fence. The sergeant there said, "Stick your barrel through that fence. I want you to fire three rounds of armor piercing ammo." An armor piercing shell is a shell that does not explode, but rather can pierce armor and along with tracers can be monitored for accuracy when fired. Continuing the sergeant said, "I want you to zero in to where your shell will go right into that opening." As he set up, Birch could just see the opening through his telescope mounted to the gun. When he zeroed down to get the gun barrel pointed into the opening of the tunnel it looked to him as though he would actually be firing into the mound in the field where B Company was stranded. So now he was faced with a dilemma, whether to say something to the officers who were waiting for him to fire, and who outranked him, or whether to keep quiet. Corcoran, his loader, could tell what was bothering him but was careful himself to keep his mouth closed and say nothing during all of this. Judging it better to point out a potential danger rather than to remain silent, Cpl. Birch decided to express his concern. Speaking to the sergeant he said, "I'll be firing into my own troops," and upon hearing this the lieutenant standing near said, "You fire! You aim for that hole." He still did not think the shell would go into the tunnel opening; he thought it would go right into B Company. He had expressed his informed concern, was ordered to fire, and so after Corcoran loaded the gun and tapped him on the shoulder, he fired.

The very first shell missed the rise in the field and landed a little bit to the right of the tunnel opening. Then the lieutenant told him what degree adjustments to make on the gun, since he and others were spotting the entire time with binoculars, while the others looked on. The lieutenant then told him how much to give it and to "Fire another round when ready!" He fired another shell. Then with some final minute adjustments the third shell went right into the tunnel opening. The slight cold sweat that had earlier broken out on Birch's forehead before he fired the first shell was now gone. He felt more at ease. Then the lieutenant said, "I want you to fire three high explosive (HE) shells. Don't touch it, fire three right into that opening." Then he and Corcoran fired three shells in succession into the tunnel opening. When the smoke finally cleared, B Company was able to make their way over to clean out the emplacement across the field. Whoever was left alive and able to walk came out of the bunker and tunnels holding their hands up. For this action Birch was awarded the Bronze Star, for performing the action out of his line of regular duty. It was beyond his normally assigned tasks, as he was designated as a tank fighter. He had joined many other U.S. soldiers and sailors who were listed as recipients of the Bronze Star. He was part of the greater team of men who had been awarded the fourth highest individual military medal. On this day he had done this, and the recognition was for his individual effort and contribution. With this single attack using an anti-tank gun and high explosive shells he had surely killed more Germans than he would in any one of his other skirmishes. In fact, when he stopped to think about it, he realized that all his tank battles combined would not equal this one act, which occurred at a distance where he could not even see the enemy's face or the direct, immediate effect of his weapon of choice.

Those who went over to inspect the scene told him, "Birch, that armor piercing high explosive ammo really put a load of them down. Did you go over to see? "No, I did not make it over," he replied. They continued to say that it was something else to see. Someone even noted that without those three shots it might have taken hundreds of men on both sides battling until the strongest side won, and "Who knows what loss of life there would be in that case? He thought about making a trek across the field to see, but for some reason decided against it. It was not that he was opposed to fighting in the war or even killing the enemy, something necessary for winning the war. At that moment, when he heard large numbers had been killed by this attack using high explosives, it somehow seemed unnecessary to travel across the field just to ogle the dead bodies of German soldiers. It sounded like something the Nazis might do, but not something he would do. He was an American soldier. He had become an expert at using his 57mm gun which was designed for fighting tanks. Yet here in one fell swoop he had taken out more individual German soldiers than he would have had he been able to destroy fifty tanks. Somehow, he felt just a little bit closer to getting home and that seemed to be satisfaction enough.

7. A Tiger Near Aldenhoven

As the war progressed in Germany, the men of the 175th continued to fight their way through The Rhineland, edging ever closer to Cologne. They fought the Germans over town after town. Their progress was slow but steady, and it felt to the regimental battalions as though they only barely crept along. He had no way of knowing the validity of the story, but Birch was told, "The battalion just received information about one little town that changed hands three times during a single day! As the evening grew late we were able to take it again. Can you imagine that? Back and forth, back and forth, back and forth, all day long. They said the name of the town is Aldenhoven." About that time, in the latter half of November, orders were issued for the 1st Battalion to move further up. The location where they were headed was this same small village he had heard of earlier. This is how Birch, and his unit, came to be introduced to this little hamlet which sat about 75 kilometers (46.6 miles) due west of Cologne. Aldenhoven was an old mining town dating all the way back to the ancient era. The Romans used to come through the area on their way toward Cologne. A late medieval water castle, that is a castle with a moat, named Scloss Dürboslar, still stood in the village with a tower that was added sometime during the 17th century, and seemed to provide the castle with a menacing yet enticing appearance. In addition to this castle, other remnants of structures from the 13th through the 16th centuries still stood in various states of wear around the village. It was for this old hamlet, where it seemed one could reach out and touch history, that the Americans now fought and secured multiple times.

The fighting between both sides was bitter and the American wounded had to be hauled out using jeeps. It seemed to the U.S. troops that about as soon as they took control of the town not more than an hour or so would pass before they heard the order, "Move back, move back," because the Germans had counter attacked and were retaking control of the town. Once the 175th had been able to establish a firm hold on Aldenhoven, Birch and his crew, along with the other two anti-tank crews, moved their guns into place. After searching for the best location to place the three-pointed trap, they decided it was best to place the guns out in a field. Their strategy was to identify the best routes tanks would have to use for travel, and one specific field provided the only viable place a tank could make its way to the town. The hamlet of Aldenhoven itself sat a little low and had a high bank onto which the men had to climb to reach the adjoining fields. From the elevated fields Birch could stand, pan the area, and see the tops of the houses and other buildings down below. About two or three doors down from the area where they sat their tank gun, they established a first aid post in one of the houses. When setting up the tank guns they typically placed them about a thousand yards apart. The configuration essentially had two guns at the corners of one side of a triangle, the base of the triangle, and then one gun further back to form the third point. On Thanksgiving Day, Thursday November 23, the unit's sergeant, who was Italian and a very good cook, made the promise, "I'm going to see that my men have hot food on Thanksgiving Day." All the men in his unit agreed that the sergeant, Sergeant Smatana, commanded his kitchen just about perfectly. Accordingly, they were especially looking forward to the promised hot meal he would prepare for them. Thanksgiving was a special treat, given that they were on the front lines with the German infantry and armored divisions roaming about the immediate area. The only way to realistically celebrate Thanksgiving was to do so with one eye on the potential routes the Germans might take in their efforts to launch yet another counterattack. Therefore, they ate with their guns set out in the field, rotating two men on point at a time who were able to operate the gun in the event a tank attempted to come through the area. By doing this, they assured their guns were always operational and ready to engage a tank. The men regarded Sergeant Smatana as the best and came to appreciate his diligence in his efforts to keep his unit fed with a hot meal "…on this day of all days." As they all enjoyed a hot meal along with Smatana, they ate with the hope and prayer that they would all have many hot meals to follow, but in peace rather than war.

The ten men of a tank crew had to be able to assure there were always troops manning the gun at any given time. They worked this out by digging a foxhole big enough for two men, and then continually rotating two men on. In addition, they strung telephone wire to provide communication capability between all three of the gun crews, as well as the three group leaders staying in the village. Birch's crew arranged to develop cover for their foxhole using the surrounding materials such that any one of the men would be able to jump over the earthen fence of mounded dirt and rocks sown with grass, and then climb down off the high bank so he could go directly to the first-aid house. This configuration provided one way in and out of the gun emplacement that could not be seen by a tanker entering the area. They stationed most of their crew down in the first-aid house and left two men on the gun, and then rotated the men manning the gun every two hours. Should the men manning the gun sight a tank, their standing instructions were to 1) communicate with the two other gun crews to develop and coordinate a viable plan of action, and 2) communicate with their group leader to keep him abreast of their status and to receive any orders he may give depending on the situation. At this point in time Birch was the group leader for his gun crew. As a group leader he always made sure to take his turn with the gun so that more of the other men were able to get a little rest. He did not want to ask the men to do anything he would not do or had not done. This was the configuration and procedure that they used for the days they were able to remain in Aldenhoven.

Two days after Thanksgiving, on Saturday November 25, the low rumble of an engine along with the clatter and rattle of a moving tank tread could be heard. Those manning the guns in the field could make out the sound some distance away, and as they did so they waited with anticipation to see what kind of tank was headed their way. Soon enough a Tiger tank came into the immediate area where the anti-tank gun crews had set up. It pulled up to where it was essentially directly in front of the gun operated by Birch's crew, though it was still some distance away. The two young men on the gun at the time were both recent replacements. This turned out to be a mistake, as replacements were always a challenge for the others in the group until they got to know the new troops well enough to be able to trust them, and better yet to anticipate them on the field of battle. With the turnovers that the 175th was experiencing, while the numbers of men could be replaced, that did not necessarily equate to replacing the capabilities of a unit. When they looked at the Tiger tank the two replacements were seeing their first German tank up close and in person.

As the tank rumbled into sight and pulled up in front of them, they both stared in shock, then panicked, left the gun, and came down to the first aid post. Leaving their post was their first serious offense. In addition to this neither one of them bothered to call Birch, their group leader, nor did they attempt to coordinate anything with the other anti-tank gun crews. This refusal to follow their standing instructions was their second serious offense. They suddenly burst into the house breathless and said, "There's a tank up there!" Birch immediately told them "You guys get back up to the gun and resume your post. Go now! What do you mean leaving the other two crews alone without any support?" Hearing this the gunner and loader both refused. "No sir," the gunner declared for both, "We are not heading back into that situation. That tank outguns us." Dereliction of duty was their third serious offense. To this Birch replied, "Well, you guys can be court-martialed for that!" With so many troops being frequently replaced, this is exactly the situation he most feared and yet wanted most to avoid.

After this exchange, he took off to keep from leaving the two other gun crews without full strength. One of his men from New York by the name of Corcoran jumped up and said, "I'll go with you Birch." This was a welcome change to the dereliction of duty just witnessed. Up to this point Birch and Corcoran had never really clicked very well. It is not that there were serious issues, but their personalities were quite different. Birch had always regarded Corcoran as something of a smart mouth, who either always had too much to say about himself or could not tell a story without trying to impress others with his vocabulary of curse words. Now it was Corcoran who first volunteered to go with his group leader. The gun required two men to operate, a loader and a gunner. The loader's job was to place the shell into the gun, make sure everything behind the gun barrel was clear, and then tap the gunner on the shoulder which was the signal that everything was all clear to fire. The two of them ran up, crawled over the fence, and then down into the placement where they had their gun dug in.

Settling into position they peered out over the gun and there sat the tank. It was a Tiger tank, a picture-perfect target that Birch could see himself blowing up that day. At the same time, going through his mind was the fact that there were three machine gun barrels sticking out of the tank, any of which was a deadly weapon. When the tanker swung the turret around Birch could see the protruding machine gun barrels. Naturally the thought came to him, "If he gets a glimpse of this gun in any way, well then he could just mow us down."

"Anytime you are ready Corcoran," Birch said, and Corcoran prepared to shove a shell into the gun, after which he would tap Birch on the shoulder once he made sure that he had cleared his arm out of the way of the gun barrel. That was the only way Birch could know that the gun was loaded, his loader was safe, and he could fire. The shoulder tap was the last thing to occur before he pulled the trigger.

Birch kept his sights on the tank, watching every move it made. As he waited for his loader, he talked to the other gun crews to coordinate a plan of action. The other two crews were further off from the tank, and so the plan they decided on was for the two more distant crews to attempt to get the tank pointed their way to give Birch an opportunity to make a rear shot at the tank. A shot from behind was the best and the easiest way to knock out the Tiger tank. Its armor was too thick from the front to be able to knock it out of commission. The other gun crews waited until the time was right for one of them to fire. When one of the other crew's 57 mm fired, the tank gunner swung his turret in that general direction, but he did not quite swing all the way around enough to expose the rear of the tank to Birch's gun. The tanker's hesitation was likely due to trying to decide whether the next shot would be from a second gun if he did swing all the way around. The situation was tense, and Birch felt a little trigger happy since this was the best opportunity, he had yet been given to destroy a tank, and a Tiger tank at that. Before the tank swung on around, toward the other 57 mm gun, Birch thought he had a good angle for taking a shot that would disable or possibly destroy the tank. With the tap from Corcoran, he fired as quickly as he could in hopes he would catch the tank in a vulnerable position before the tanker had the opportunity to swing back around.

The 57mm anti-tank gun creates a large muzzle flash when it is fired. Because of the proximity of the tank to Birch's gun, someone in the tank got a glimpse of the flash from his gun and the tanker began to swing his turret back around in the general direction of the flash. At this point the 57 mm gun and Tiger tank were probably within about 61 meters (200 feet) of each other. When the tank started to swing back toward their gun, Corcoran threw in another shell and punched Birch on the shoulder. Wanting to get another shell off before the tank had the chance to turn all the way back around so it was facing them, Birch fired his second shot. When the tap was given, it meant everything was a go to fire, but it turns out that Corcoran had not gotten his arm back out of the way before he punched his gunner on the shoulder. The 57 mm gun had a recoil of three feet which was enough of a kick to break his loader's arm when Birch pulled the trigger. Of course, with a broken arm Corcoran could not do anything else to help with the battle; he was out of commission. "Get out of here Corcoran or he will mow you down for sure, but you have to hurry, you have to go now!" Birch shouted to his loader. "What a stupid mistake," moaned Corcoran, "the pain is killing me!" Birch replied, "Go now, now before he has a chance to zero in on you!" Groaning in pain, Corcoran made one awkward attempt to haul himself over the earthen fence using his one good arm. Making it over, then, he could roll or crawl down over the slope where others could help him. Birch would have helped his loader up and over except that he was too occupied with the Tiger to notice whether Corcoran made it or not. From his perspective behind the gun, he could see that the Tiger was not done searching or shooting.

After searching in the general direction from which the previous shot came, the tanker then took aim and fired his turret because he could still see the residual smoke as evidence of the anti-tank gun's location. Though the tanker did not know exactly where Birch himself was located, he had a general idea and was working on narrowing this down. The last thing Birch remembered was looking into the barrel of the turret gun on the Tiger tank. This was the tank that made such a horrible shrieking noise when fired that he felt like he was going to come out of his skin. He absolutely hated to hear the screeching scream of the Tiger 88s when they went off and considered that the Tiger with its artillery might very well be the most dreadful weapon in the German arsenal. Now this horrific weapon was no more than 200 feet from him with its artillery pointed directly in his direction. He knew he was looking at certain death if the tanker decided to fire, and he could do nothing about that.

Time seemed to slow, when suddenly there was a horrible shrieking noise and Birch felt as if he were briefly floating. Suddenly, unexpectedly he was struck in his back by a ferocious force. As he regained some of his senses, he was a bit dazed and confused at the situation. He had expected a force from the front, not from his back. This made him wonder if there was another

German attack from his rear position. In fact, what had just occurred was that when fired, the tank's shell hit close enough to him that the explosive concussion knocked Birch away from his gun back about twelve feet, where he landed on a pile of hard-packed dirt. This explained his brief sensation of floating and then receiving a hard blow from behind. His helmet strap remained buckled throughout all this. As feeling began to return to him he felt as though the whole ordeal just about broke his neck. He could hardly move his head at all without a great amount of pain in his neck. He had hit so hard that his back was painful and sore, and his throat was extremely raw and painful from ingesting a large amount of dirt and fumes when the shell exploded beside him. Just then the thought came to him, "If I raise up, he is going to machine gun me. But I cannot just lay here because he is still searching for me!" Still dazed he felt the weight of the quandary in his mind whether to move or not.

As his head cleared a little, he knew he had to move away from his gun because he could see the tanker searching, with the turret moving back and forth like some gigantic mechanical eye looking for any sign of movement. For Birch the tank seemed like a gargantuan entity that roamed about, looked for prey, and then attacked at will. As part of his training on the 57mm anti-tank gun he and his fellow soldiers rode in tanks; they trained in tanks. He knew what those inside the tank could see, and that their vision was limited. The tankers had a vision port that was approximately three inches wide and six inches long. Two men normally rode inside and when they were closed up, inside the tank with their top hatch closed, then all they were able to see were those things clearly visible through this slot. That is what they had for observing and searching when the tank was sealed up and moving. If a soldier could hide, he could let a tank come right up to him and then he could stand up and walk along side of it. There would be no way for the tanker to know anyone was there. The only thing was, if the tanker happened to have some buddies stationed over some distance away from the tank they would know.

Birch knew that those inside the tank could not see much, and with the limited experience he had riding in the tanks he knew that the tanker had just that slot to look through. Therefore, he figured that he had a chance, one chance, to make one and only one lunge over the fence. For him to be able to leave the gun, he would have to make it up and over the earthen fence to have any chance at all to put some distance between himself and the tank. The fence was his only way out without being seen, but he knew he had better go and go right then or else the tank was going to zero in on him, or close enough to him, to end his life. He thought, "If I can get over that fence before he is able to pull that trigger on the machine gun, I might have a chance." Unknown to him was that he no longer had full strength in his legs. Laying there on the dirt pile he was in pain, but he assumed all along that he had his full leg strength. Yet when he rose up to make his jump, he discovered he could not walk, at least not with one bad leg. Using the strength of his good leg he barely made it up onto the fence, and then was able to push himself on over so that he could continue to roll down the slope where some of the men ran out to retrieve him, and then took him on down to headquarters. At headquarters he laid there, and a soldier from Middletown, Ohio happened to be there that night. The reason this was notable to Birch was that Middletown was the area in southwest Ohio where he himself was living with his family when he entered the war. This young man from Ohio and others there patched him up for the trip that he would need make to the field hospital. For the time being Birch's part in this war stopped. He needed to be taken off the field of battle so others could better tend to his wounds.

Soon the Germans sealed off Aldenhoven again and the Americans could not take their wounded out. They had been able to patch Birch up, and during this he heard that he was not injured as badly as some of the other men. It seemed amazing to him that, yet another foul up led to his and Corcoran's injuries. Only this time the foul up was on the part of two replacements who blatantly committed three military offenses in a period of something like five minutes. They could not really cover up their offenses as they committed them in full view and hearing of the other soldiers sitting there in the house. It had to have unsettled and angered the men who faithfully served day in and day out to hear the two refuse to do their job, as well as refuse a direct order. Had they coordinated with the other gun crews and then called Birch, things could very well have gone differently. Instead, he and Corcoran manned the gun and later both ended up headed to a field hospital. It was also overwhelming to him to realize he had been shot by the turret artillery on a German Tiger tank and yet to find that there were others worse off than him. He was going to have to heal but was beginning to feel more confident that he might actually live through this to talk about it. About that time the young man from Ohio said to him, "Sir, we don't have any place to put anybody else, we are full and using every available space as it is. We'll have to put you back in this potato bin." "Well," replied Birch "I am pretty comfortable with potatoes, and would much rather be here, than laying out there in the field." His first hospital bed, then, would be with the potatoes. He laid there and thought about the irony of starting out his tour of duty by serving on KP duty on the RMS *Queen Mary*, and that now he was back with the potatoes, laying there by them as though he were just another part of the pile. He laid there in the potato bin until the other troops could manage a way to get him out. Laying there he wondered whether the Tiger was able to go on through or whether it was taken out by one of the other two remaining gun crews. If only he could find out.

The decision was eventually made to attempt to get the wounded out of the area after nightfall. This was going to be their best chance, but it turned out that it was long after midnight before they finally made the effort to move him and the other wounded. The countryside and woods made it very dark, enough so that the vehicle drivers would not be able to see the road, and of course they could not use the vehicles' lights with the enemy nearby. Part of what took so long was that some method had to be devised to help with the evacuation of the wounded, many worse off than others. Time was of prime importance for this medical mission, which was becoming more urgent and more complicated by the minute. What they finally decided on was to clear a way that could be used for sneaking the wounded out. At least they figured they would be able to sneak them out. The wounded were loaded up onto jeeps, and then white cloths were given to men who volunteered to act as guides. Earlier the Germans bombed the roads, which created bomb craters deep enough so that the drivers had to work their way around them. The craters were too deep and treacherous for the jeeps to be driven over or into them. The guides leading the jeeps out used white cloths, white handkerchiefs, or whatever else they could find. The guide got in front of a jeep and led it out of the town back far enough behind the front line to where it could go on its own. They said nothing out loud but rather used hand signals along with the white material to indicate which way the jeep drivers should steer. In this way they led the wounded out to safety.

They traveled back behind the front line, but not very far back, to where they had a field hospital. The field hospital, located a little distance behind the line of battle, was where they operated on Birch that night. The reason he was unable to walk was that one of his legs and hip contained quite a bit of shrapnel from the exploding shell from the German Tiger tank. That night they operated to remove the steel out of his knee and hip. He did not remember anything much after that and was not even certain how long he remained in the field hospital. He was not recalling much or keeping very good track of time as he was in and out of consciousness due to the medications. As it turns out, from the field hospital the next morning they loaded him into an army ambulance and took him further back to a Belgian hospital. He stayed there for a longer period under medication where Belgian nurses who could not speak English cared for him. The language barrier really meant nothing at all regarding the care he received, for they seemed to know exactly what he needed when he needed it. To his way of thinking he had never received any better treatment than he did from these Belgian nurses. Though he noticed that he was starting to heal, his throat remained very sore for two weeks after the attack. It felt to him as though it would never be the same again.

One day as he lay in his hospital bed a couple of soldiers came through the area and examined the tags that had been placed on the ankles and feet of the wounded. Earlier the medical personnel identified the patients but did not tag the beds. Rather they placed a tag on the patients' ankles. As they made their way through his area the two soldiers picked up Birch's tag, looked at it and said, "Yeah this is him." Then they loaded him up into another ambulance without saying a word to him about his destination. At this point he was becoming a little confused since he had no idea where he was headed. He noticed that on the way that he kept hearing train whistles. After a while they pulled into a place he could not recognize, sat there for what seemed to him like quite a long time, and eventually unloaded him from the ambulance and put him onto a train. It turns out that the train he was loaded onto was a hospital train and before long he ended up in Paris. For the first time he found himself in what he determined to be "a real fancy hospital," and his room was located on the ninth floor. The hospital, the city, the view out of his window, which included the Eiffel Tower, were all beautiful. The Americans had taken over the hospital and American nurses and doctors were there to take care of him along with many other U.S. soldiers who were gathered there. He remained in the Paris hospital until December 30, just in time to celebrate the New Year.

On the 1st Battalion's daily report from November 25, it simply stated that Cpl. Birch and Pfc. Corcoran were moved from active duty to the hospital due to being wounded in action. There was no elaboration or acknowledgment as to how they were wounded. That is how it always was with these reports. With abbreviations it stated things as simply and as blandly as possible. For Birch though it was not simple or bland. It was much more than merely being wounded in action. On the surface of things that was true, but he had spent the past weeks not only healing from his physical injuries but battling the tank all over again in his dreams and reliving it over and over in his mind. It was bad enough the first time but reliving the experience in a dream or memory was just as hard on his emotions. The fear was real each time. It was as if emotionally he was being pounded over and over, and then over again by the Tiger artillery. Each time, he heard the unnerving shriek that he so dreaded hearing on the field of battle. Furthermore, each time his emotions were being ravaged all over again. While his body was continually healing, he felt as though he had to emotionally heal all over again each and every time he faced the Tiger in his mind. The entire time he was in the potato bin and subsequently in three hospitals for surgery and time to heal and recover he had hoped he would get to go home, back to the United States. Many injuries sent men home and his was certainly not a minor injury. It came about though that he was not to head west back to the States, but east back to his unit. On New Year's Eve the 1st Battalion daily report listed him as coming back to duty on December 30 from the replacement depot, meaning that he rejoined his unit two days before New Year's Day, five weeks to the day that he was shot by the Tiger tank. Moreover, when he came back to his unit, he found that it was very nearly in the same spot it was on the day when he was shot by the tank. He would probably be able to find out whether the tank had been destroyed, or whether it made its way on through to its destination.

The process of taking him back to his battalion involved loading him onto a truck, taking him to the train station, and then putting him on a train and sending him back up to his unit. From the train station on the other end, he was driven again by truck back to the kitchen area of his battalion. The kitchen area of each unit always remained back about five to ten miles from the main front lines. From the kitchen area he was supposed to go on up by jeep. Something different had occurred during his five weeks away in that the men seemed to be having a field day when he returned. They were out shooting into the air that day, just as if it was a celebration. He was told that Germany essentially sent every remaining plane they had out that day trying to make what amounted to the Luftwaffe's last stand. He saw men out on the fields of battle and the thought came to his mind that "They are out there as though they are having target practice on balloons or something easy enough to shoot." It seemed to him that they were firing at the planes and shooting them down as though it was no problem at all. Planes were falling from the air, and everybody was vying for a chance to fire, "Let me shoot" one would say. Everybody wanted his chance. They were out in the fields laughing and at this point having themselves a ball. Though the war was not over, this was essentially the end of the German Air Force; the Americans had destroyed the Luftwaffe that day. When he finally got back up to his unit, they had not been given the typical tough assignments to which they had become accustomed on the front lines, primarily because they were anticipating being able to go on in to take Berlin.

One thing that Birch finally learned after he returned to the same area where he had been shot was that one of the other gun crews was able to take out the Tiger tank that day back in November. One day in January 1945 as he was walking through the area, he spotted a tank, the tank, the Tiger tank sitting in the same spot that it was when it fired on him and blew his world up and nearly killed him. Only something was different about the tank from five weeks past; it looked old as though it had been sitting there for years. He was a little amused that the entire tank was rusted out due to the fire that erupted from the explosion when the tank was shot by the other gun. This is the reason it appeared as though it had been there for years rather than weeks. While that day in November the tank had the advantage over him, it made no further progress from that point on. All three anti-tank gun crews worked together to overpower the Tiger. It was hard for him to imagine that anyone in that tank could have survived the shelling, but then men seemed to survive some of the fiercest exchanges of the war, as he felt he had. Still yet, it also seemed true that men had been killed by fleeting exchanges that occurred in an instant, like what had happened to his former gunner Jimmy, who died in the time it took for Birch to hear four machine gun rounds thump into his chest. In his mind he thought, "Isn't it odd that men in the field of battle can be so durable and robust and yet so fragile at the same time?" He briefly wondered if the tanker who tried to kill him had survived the attack on his own tank. Then he walked on past the rusty vehicle now sitting beneath the bare limbs that earlier held the bright fall foliage so prevalent on the day he met up with the Tiger.

8. Papa No Come Home

FURTHER EAST FROM the area around Aldenhoven where Birch had earlier met up with the tank and been wounded, the 1st Battalion approached to within a short distance of the city of Cologne. The battalion's daily report of February 27 read "Attacked and captured town of Borschemich today. Complete operation took 4 hours with very good cooperation from tankers." Now, in early March of 1945, they were waiting between Mönchengladbach and Cologne to progress further east after crossing the Rhine River. Birch stood up on a hill across from the Rhine River some distance from Cologne and watched as the city was systematically destroyed by Allied bombers in air raids over two days. These bombing runs were essentially the final blows to the city after three years of bombing.

Cologne had already been the target of 262 separate air raids dating back to May of 1942. Most of these bombing runs were performed by the Royal Air Force (RAF), with the U.S. Army Air Force (USAAF) joining in the early days of March of 1945. The largest single raid came on May 30, 1942, when over 1,000 bombers were sent to make a nighttime bombing of both chemical and machine tool factories in the city. Now, from the men's perspective on the hill, both the USAAF and RAF bombers laid the city "flat as a biscuit," as Birch expressed it. This aggressive effort was due in large part to the fact that Cologne continued to be regarded as an area of high resistance. These last RAF bombing raids on the city, witnessed by the members of the 1st Battalion watching from the distant hill, were designed to occur in waves and involved a total of 858 aircraft. It was a daylight bombing run designed to be highly destructive. Birch and his unit sat up on the hill overlooking the area around Cologne since they were again waiting, unable at the time to get across the Rhine River.

The Germans had blown up all the bridges on the Rhine, and in March of 1945 it was cold. In addition, the snow in Germany that year got as deep as four feet. Everything was made more difficult due to the cold, deep snow, whether it was fighting or waiting. The harsh German winters could quickly become another enemy to the soldiers in the field, especially those who were unaccustomed to them. It seemed that every time the U.S. engineers worked on the construction of a pontoon bridge the Germans merely looked on, patiently watched from a distance, and allowed them to complete the building of the bridge across the wide Rhine River. Then, just as soon as the U.S. troops completed the bridge over the Rhine, the Germans blew it up. Birch and his unit sat there for a few days waiting until they could somehow cross the Rhine. They sat on the hill, watched the Allied bombers fly overhead and bomb the city of Cologne, and waited for the sign that they could make the river crossing.

Just a short distance from Cologne were several smaller towns, some deemed more crucial than others for the Allies' plans to defeat Germany. Aldenhoven and Eschweiler were among those smaller towns that had robust fortifications that needed to be destroyed by Allied bombing raids. Other small villages such as Erkelenz, and Borschemich were secondary targets that were also regarded as crucial small towns. For the most part these were hamlets of less than three to five thousand people, if they were that large. The two sides fought fiercely for those towns, and it happened that Birch and his unit sat near Borschemich, just outside of Erkelenz, from March 11 through March 31. Since there was no tank action occurring in the immediate area at the time, he and his crew, along with the other two 57mm anti-tank crews, sat back and watched the war efforts continue to be put into operation around them. They remained out of rifle fire, but not out of artillery range because they were still on the front line and ready at any moment to set up their 57mm gun configuration if, and when needed. For this short period, though they were not called on to fight tanks, they still were on call every minute of every day. They took this opportunity to sit back, talk and laugh, and await their orders to move up. Until this order came, they would have to wait a means for crossing the Rhine.

They were always careful to maintain contact with the command post through their officers, who were the sergeants and lieutenants. Since their landing on Omaha Beach, one problem they could not seem to solve was they were never able to keep a lieutenant for very long. It seemed that about the time they began to get used to a lieutenant the Germans would kill him. Therefore, for most of the time their sergeant served as their immediate officer in charge. Their unit had a sergeant from Maryland, Sergeant Long, a farmer by occupation. Long was a short guy who did not seem to care for anything, in that not much seemed to shake him or distract him from his duties. Corporal Birch and Sergeant Long started buddying together ever since they were back in England training for the invasion at Normandy. Later, after the 1st Battalion landed at Omaha and started to move inland, the unit had two lieutenants who were killed one right after the other. Their original lieutenant had been killed shortly after they landed on shore back in June of 1944.

After they lost the two lieutenants in succession, subsequently another one was sent to them from the available pool of officers. The men of Headquarters Company were not certain whether this new lieutenant was already in Europe or had come to the field directly from the States. Either way it was apparent that he was fresh out of officer's school and bent on doing everything strictly by the book. The problem was that life on the front line in the middle of the war did not always allow for operating by the same book. This was a luxury of the troops stationed at home on a military base during peacetime. The new lieutenant came to the front line with one book and was bent on making sure that his unit obeyed U.S. Army rules to the letter. He was committed to making sure that the men shaved every day and that they saluted properly. Sergeant Long, out of concern for his lieutenant's safety, as well as that of the men, told him, "You don't know what you're talking about. We don't have no rules like that over here. You're going to make your men salute you, and you're not going to live a week." He had not said this out of disrespect but rather because officers in the field, especially on the front line, took the bars off their uniforms.

An officer was not supposed to wear bars on his person because that gave the enemy a sure sign he was an authority, a leader, and the enemy strategy was to kill the leaders first and foremost. If the leadership was taken down and out, the rest of the troops at best would require more time to organize, plan, and develop a cohesive battle plan. In the worst case it could completely upset the troops and cause chaos. To let the men in the field know who the officers were, each officer placed a white strip of tape on the back of his helmet. This white piece of tape was oriented horizontally if the soldier was a noncommissioned officer, that is, if he was a corporal or a sergeant. On the other hand, the piece of tape was oriented vertically if he was a lieutenant, or if he was an officer of some higher rank. That was the protocol for how the men identified their leaders, by the tape on their helmets rather than by bars on their uniforms. It was also protocol on the battlefield not to salute officers. On the field of battle in France and Germany the enemy was very near most all the time, and able to watch every move the Allied troops made. Saluting an officer was as good as putting a target on him, and many units had seen their officers targeted and killed by such thoughtless acts.

Given all of this, the new lieutenant was still determined to have his men behave like American troops when on military bases back in the States. While the lieutenant planned to have them shave every morning, the men on the front line, during fighting, would typically go for a week and not shave. In addition, the soldiers in the units wore dirty uniforms, and did not have the luxury of keeping a clean uniform and clean boots each day. They were soldiers in the field, on the front line, fighting the enemy every day, and sometimes more often than that. Moreover, they lived out in the open and under some of the worst weather conditions, making it difficult to stay dry or warm, let alone clean inside and out. This new lieutenant came in among them and wanted them to salute and do everything else by the book, just as he had been taught in his basic and officer training schools back in the United States. He was persistent in having his way, no matter what Sergeant Long, or anyone else for that matter, had to say by way of good advice. This is also why he was only their lieutenant for about ten days before the enemy killed him. While the men hated that this happened, they also expected that it would soon happen. No one was really surprised at all. If anything, some had even expected him to be killed before that.

As it turned out they were content to carry on and do their jobs along with Sergeant Long as their commanding officer. Long remained their immediate commander all along up until August 26 of 1944, when he was wounded and left action for a while. In their time getting acquainted with each other, Birch came to regard Long as one of the most intelligent men he knew. There seemed to be a genuine friendship between them. Long would routinely take Birch with him of a morning when he made his daily trek to the command post. Though Birch never wanted to take the time to go, Long would say, "Yeah come on Birch, go with me." He would respond, "No I don't want to go" and Sergeant Long would answer, "Yeah, I may have to send you back after the crew." Birch would reply, "No" and Long would counter, "You don't have to worry, its them bullets, as long as you hear them bullets you don't have to worry, it's the ones you can't hear that will get you," to which Birch said, "Yeah, I know that." This became their daily dialog, like a ritualistic dance they had committed themselves to complete before Birch and Long would usually hike together up to the command center.

The day came though when Long was wounded. That morning he had gone up to the command post on the front by himself; Birch had not gone with him. Long went up to get orders to determine what their unit was to do next, when he was wounded. The news of Long's wounding just about took the heart out of Birch. He became depressed because he had known the sergeant for a longer period than anyone else in his unit that he could recall, ever since they met in England. They had buddied together nearly the entire time and he regarded the sergeant to be an excellent leader who watched after his men with genuine concern. Sergeant Long made sure they were safe, as safe as they could possibly be in a war. This was not always the case with the officers in command. Compassion for one's men was a unique trait that was only characteristic of some leaders. For the first time since making footprints on European soil Birch entertained the thought about how he might wound himself and get out of the whole situation. Of course, he never seriously made a move to do something like that but that is how much he thought of Long, and how depressed he felt at the time. Long was sent to a hospital back behind the front line and was dropped from the rolls for active duty on August 26, 1944. However, while Long was down, he was not out. Like Birch after his wounding, Long eventually fought his way back to wholeness, and made his way back to his old unit. The team of Long and Birch continued from then on to serve together for some time to come.

While in Borschemich, Germany the 1st Battalion of the 175th Regiment of the 29th Infantry Division attended a ceremony wherein the unit was awarded a Presidential Unit Citation which they received for their actions on Purple Heart Hill back on June 18 of 1944. The ceremony occurred on March 17 of 1945. The battalion daily report noted, "Borschemich, Germany, 17 March 1945. Ba. received Presidential Citation today at 1500 hrs. The presentation was given by Brig. General Sands. The award was accepted by Maj. J. Clarke Geiglein, Ba. C.O."

As he stood there with his unit, Birch thought with great sadness about the many men of his original unit who fought so hard on June 18 of 1944 and survived the battle in the field on Purple Heart Hill, which resulted in so much bloodshed and death. It seemed unfair somehow that men like his gunner Jimmy and many others survived being surrounded and outgunned on Purple Heart Hill only to be subsequently killed in action. These heroes would not be able to experience the awarding of the Presidential Unit Citation that their Commanding Officers had recommended and moved forward after the unit's actions on Purple Heart Hill. He knew, as did all soldiers, that there is no fair or unfair in war, there is just what transpires. Had he not traded foxholes that night back on Purple Heart Hill he too would have been killed by the mortar shell lofted over the fence in the night. Had the German Tiger tank's shell hit just a bit closer to him back in November, just after Thanksgiving, he could very well have been carried out in pieces. No, it was not that the others were cheated, but his heart was still heavy, especially given how close he had worked with his crew since their time in France and Germany. It was always difficult to see a man go down, but in certain instances it somehow became very personal and raw. While the rolling motion of the R.M.S. *Queen Mary* served to mark off every twenty-minute period, amounting to eleven miles further from home, here on the battlefield the distance from home seemed to be measured in lives. With every life lost, every man down, Birch felt like he was moving farther and farther from home. Sometimes it was all he could do to fight off depression and the Germans at the same time.

As his unit continued their progress toward Berlin, they eventually got to within about 64 kilometers (40 miles) of the city. Their sights were set on Berlin, and they had their vehicles loaded in preparation for entering the city as soon as possible. They wanted to see Berlin; that was their heart's desire. It was as though the city itself was the grand prize, the final goal in a game that is played whereby you have to physically fight your way to the goal. Those who fight and survive are the ones who get to enter the city and win. Some of the men felt as though not making it into the city limits of Berlin would somehow be like giving something up. Therefore, they were determined that they were going to make it there. Instead, it soon became clear that they would only manage to get to the Ruhr River. They were within reach of Berlin when orders came down for them to stop at the Ruhr. So, they stopped and camped there for several days, waiting for the Russians to cross over the river. The Russians wanted to take Berlin, and to that end they wanted to meet the Americans at the Ruhr. All of this was negotiated at higher levels, beyond the reach or input of those fighting in the hamlets and fields of Germany. The soldier in the field was not a part of such decisions and arrangements that were made.

While the Americans waited at the Ruhr, more and more Germans began to appear, streaming down to the Ruhr where the American soldiers were encamped. Over time they numbered in the thousands. These multitudes requested the U.S. soldiers to take them on over to the American side of the river. It was clear that they felt desperation and would do anything possible not to be captured by the Russians. Since the Americans were waiting and did not have other assignments until after they met up with the Russians, they decided to get boats and take the German citizens across the Ruhr. This was a large undertaking and they ended up boating these citizens over the Ruhr River to the U.S. side day in and day out, until the Russians came down to meet up with the American troops. All totaled they estimated that they had taken thousands of Germans over the Ruhr to the Allied side. It was a sad and frightening sight to see German people running from the Russian military because they knew that they would be the ones to pay the price, and pay dearly, for the efforts of the German military who acted against the Russians on the command of one man, Adolf Hitler. Birch knew that it was not right that in this war between nations, it was the citizens of the countries involved who were intentionally punished, and even terrorized by the military. "It is just not right," thought Birch, "but then some of the governments involved in this war are bent on terrorizing their own citizens, and so will not give a second thought to doing the same or worse to the citizens of other countries. In their eyes these are not people with lives, just assets to be used as they wish. The people realize this, they know they are not regarded as individuals with lives and aspirations. They are regarded as possessions. This must be why we have seen thousands over the past days coming to us for help. Now I only wish that we had longer to wait." The American soldiers could see the fear and desperation in the eyes of the men, women, and children who were fleeing for their very lives. This was a mission the Americans took on but were in no way expecting until it happened to fall at their feet. It was a good thing that the troops were under no orders to move further on, but rather to wait for the Russians. The waiting gave them the time necessary to help as many German citizens as possible.

The unit in which Birch served was part of the effort to go into the towns for clean-up operations in the event German military personnel were still hiding out there. The procedure was that they would go in, then if it was not a tank area the anti-tank crews could sit idle for a bit. This was the closest thing to getting a respite in the field since they had entered Germany. Part of cleaning up had to do with searching for snipers. There were always snipers. As they went through town after town, they would nearly always find snipers waiting. In the larger cities there were typically greater numbers of snipers hiding and waiting. Headquarters Company was the company in which Birch served, and in addition to housing the tank fighters one of its assignments was military intelligence. Headquarters Company had received intelligence informing them of a German officer still lodging in Düsseldorf, one of the larger cities in Germany. Moreover, they learned that someone had located him and knew where he was living. Sergeant Long and Corporal Birch were given orders to go capture this man and arrest him. He was a high-ranking officer, or at least he was supposed to be according to the intelligence that they had received. After receiving orders, the two of them traveled to the house where they had been told the German officer was staying. A lady opened the door, and it was apparent that she knew exactly why they were there. The officer came down the steps, cleaned up and wearing a uniform, apparently his dress uniform. Long and Birch did not treat the man in any way that was offensive, but rather treated him as an officer, which was required under military law. They took him from the house, placed him under arrest, and marched him out to their jeep. Naturally they kept a gun on him but did not physically bind him in any way, say anything violent toward him, or attempt to deride him in any way. In accordance with their instructions, they took him down to intelligence and turned him over to the proper personnel there. It would have been easy, especially given the experiences of the past months, to do or say something to this officer by way of gloating or taunting, but Long and Birch both remained professional about the entire process. Still, in his mind Birch did not for one second believe that were the circumstances reversed, the German officer would have been as professional, nor would the German have refrained from making some attempt at taunting him or at the very least do his level best to make Birch uncomfortable so that his trip to wherever he was going would be as tortuous as possible.

As he thought about how others might perceive treatment by Americans, a memory came to mind regarding someone, a civilian, who had apparently been helping the enemy side. A while before they reached the Ruhr River there was a small town through which they passed. These small towns were typically only four or five miles apart, and one could usually be in one town and look across the open fields and see into the next town. Birch's unit happened to be stationed in just such a town for a few days. It was wet and muddy, cold, and very rainy. The mud was so deep in the streets that they could hardly walk. The tanks in the area tore things up half the time, including the roadways, which made it even that much worse. His unit was waiting again for their orders to continue moving. They had their guns set out along a fence and they were once again in another apple orchard behind a house. It seemed like they often ended up in an apple orchard. The guns were pointed to the other town that could be seen across the open fields. If tanks came through that area from one town to the other, then Birch and his crew would be ready. The unit had taken over a farmhouse which was occupied by a woman who had a teenage daughter. Her husband had either been killed in the service or was presently serving in the military. They did not know anything about him, but he was not living there on the farm with the two women. The women could not speak English, and the U.S. soldiers did not have anything directly to do with them. Birch and his crew arranged to share part of the house with the women, whereby the men took one section just for sleeping quarters.

Every morning Birch would go down to check the gun crew because they changed guard on the guns every two hours. Two men would come off and two men would go on. As group leader for the team, every morning Birch would go down to check on the status of things. It occurred to him one day that after he went down to the guns the Germans would shell the Americans at that the spot where they were staying. When this occurred, artillery shells would continue to rain down on them for about an hour. Two houses up from that house was another farm where an old man lived with his wife. Over time Birch noticed that this old man two doors up would go out to a chicken house located some distance from his house, and afterwards the enemy shelling would start. Accordingly, the decision was made for the soldiers to go through the old man's house and barn and search for anything that might be suspicious. The house and barn were built together with the roadway passing between them. Part of the barn was on one side of the roadway and part of it was on the other side along with the house. More than once the soldiers went through the house and barn and then out into the orchard where the man had a chicken house and a couple of other small buildings. Every time they decided to search, the old man would just smile and nod his head. Birch noticed that the old man would go out, and then later they would be shelled. One of his crew had even made the connection and then commented, "Every time he goes out, we get shelled." "I know," said Birch, "and we need to get to the bottom of this. We have checked him out, but something does not quite fit. Something is off."

Birch decided to report this to his commanding officer. When their commanding officer heard about this he said, "Check that chicken house out, and see what's in it." Therefore, one morning Birch went down and got the old man and took him outside and they went through the chicken house but could not find anything that was clearly a problem. He did notice, however, that there was an awful lot of wire laying around in the building. He also noticed that it was not merely one type of wire, but rather all kinds of wire being stored there. On the other hand, he and the others found nothing there in the way of any kind of broadcast equipment or accessories. Therefore, he let the man go and reported back to the commanding officer. The commanding officer said, "Arrest him, bring him down and interrogate him." On the next morning he made another trip to the old man's house, knocked on the door and the old man came to the door. Birch did not speak German and the old man could not speak English, so Birch motioned for him to come with him. He had taken another guy with him and had instructed him, "Now I am going to take off down the road and he's going to follow me, you stay behind with the gun." After Birch motioned for him to get ready and follow him, Birch put on his cap and started off down the road with his prisoner following right behind. He took the old man to headquarters and turned him over to the soldiers there. Their unit was there in the little town for three more days. While there he continued his routine every morning, of going down to check on the guns. The woman of the farmhouse knew he was the one that came and got the old man, and Birch felt somewhat sorry for her. When he made his rounds each morning she would come outside, use her apron to wipe her eyes, and cry out saying, "Papa no come home, papa no come home." Birch just kept walking because there was nothing that he could tell her, and due to the language barrier, he would have had trouble talking to her even if he did know something. Each morning she would plead her case, "Papa no come home" and then the same the next morning, and the morning after that. She would wait until she saw Birch coming down, and then she would go out to where he was and say, "Papa no come home." It seemed to be her way of pleading her case with him, but he had only picked the man up and taken him to intelligence personnel for questioning. Other than that, he knew nothing and was not brought into the process after following his orders. Some days later, when their unit pulled out of town, papa still had not come home, and Birch never knew what was done with him. He left feeling sorry for the woman whose husband left and had not yet returned. Though he could not communicate with her due to the language barrier Birch was somehow chosen by her to be her sounding board, someone to whom she could appeal. Now he had left the town and she had no news, apparently nowhere to go, or anyone to talk to about papa.

9. The Barn

WHILE ANTI-GERMAN sentiment arose in the United States during the Second World War, it did not quite present itself as powerfully as it had during the First World War. German American loyalty was not as much in question as it had been earlier on. Dwight Eisenhower, a descendant of the Pennsylvania Dutch, was in command of the U.S. troops in Europe and the industrial expansion which occurred during the Second World War also helped to bolster the assimilation of many German Americans into American society. After the war's end even more German immigrants came to the United States. In large measure these consisted of German people hoping to escape the aftermath of the war. These new arrivals were extremely diverse in their political viewpoints, financial statuses, religious beliefs, and they settled throughout the United States.[1] These added to the numbers of German immigrants arriving in the United States during the opening years of the war as they fled Nazi persecution in Germany which so many varieties of people suffered.

In the early weeks of the invasion of Normandy, Headquarters Company of the 1st Battalion included a young soldier who was a native-born German. While he was still living in Germany this man's parents were killed by the Nazi party which recently rose to power. Since the soldier did not talk to just anybody of his past, most of the men in Headquarters Company were uncertain regarding whether or not this German American soldier was Jewish. Nor were they clear as to whether his parents' deaths had to do with them being Jewish. Those who knew him best knew more details of the tragedy and tended to be protective of him. No one else wanted to pry. Furthermore, while the details were not clear to his fellow soldiers, they all were aware that his parents had been killed by the

Nazis. Somehow, through his family connections, this man made his way to the United States and joined the U.S. Army. This was likely facilitated by the fact that the U.S. Congress provided for the expedited naturalization of non-citizens who served honorably in the U.S. Armed Forces. This German American intentionally joined the Army for the purpose of getting into an intelligence unit as a translator. Since Headquarters Company was an intelligence unit for the 1st Battalion, this happened to be the unit to which he was assigned. Translation was a very natural fit for him and an easy fit for him once he was in the armed services.

He was exceptionally good as a translator because as a natural-born German citizen he was intimately familiar with various nuances of the language, as well as differences between dialects throughout his native country. At the time, in accordance with their orders to capture at least one prisoner every day, the company's efforts paid off by way of the capture of a sniper they found hidden in a tree. Soldiers from another company of the 1st Battalion spotted him, disarmed him, and then forced him down. They brought him to Headquarters Company and asked the unit about obtaining a translator. The German American soldier happened to be nearby and upon hearing the conversation answered, "I can translate for you." Little did anyone else around know that his offer to translate would soon lead to a hostile, near fatal, encounter.

When they brought the sniper to the translator, the translator's attitude suddenly changed. The translator only just began speaking to the German sniper in his native tongue when the other soldiers suddenly found themselves having to restrain the translator from physically attacking the sniper. They could barely hold him back to prevent him from doing the sniper bodily harm. Whatever exchange occurred between the two of them in German angered the translator. He quickly lost his interest in interrogating the sniper; he now wanted to kill him. The intervention of the others prevented him from slitting the German sniper's throat. For the translator, the dialog became very personal rather than remaining a routine mission assignment to obtain information. Within just a few minutes his fellow soldiers finally helped him calm down a little, after which they chided him saying, "No! We want information from him, to know what's around. You're not going to kill him."

When the interview continued the translator was able to make the sniper answer quickly and tell them what they needed to know. He had a knife, the sniper could see that he had it, and could sense that the translator would have no hesitation to use his knife to finish him off right then and there. For everyone present tension and anger was felt, like a dark fog hovering in the air. The translator spoke to the sniper without pleasantries or patience. Instead, he spoke to the sniper in harsh, accusatory tones. Looking on Birch thought "It is very likely that this German prisoner is under the impression that if he does not tell us what we want to know we will allow the translator to have his way with him. For all anyone present here knows, the translator might have communicated that message to the sniper." For whatever reason, the sniper quite willingly communicated to the Americans in a cooperative, talkative tone.

Those looking on seemed to feel that, given the opportunity, the translator would personally have knifed every German soldier he came across. His sentiments toward the Germans were understandable to his fellow soldiers, and many of them even empathized with him to the point they would have helped him do the killing by hand. What struck them all as amazing was his advanced planning to come to the States, join the United States Army, and get into intelligence where he could have direct access to captured German troops. It was an elaborate plan. It was not so much that he joined the Army under false pretenses with a hidden agenda, as it was that when he directly encountered the German soldiers all the deep-seated emotions from whatever had been done to his family came flooding back to him, making it difficult, even almost impossible at times, for him to maintain control over his feelings and desire for revenge. In a sense, his overwhelming emotions which he was unable to hide made him a more effective interrogator. This seemed to work in his favor with the sniper, who was clearly afraid of him more than anyone or anything else, even before the interview got fully underway. If he could make other German prisoners feel this way, he would most likely start off every interview with the advantage. The translator's situation, even without all the details, tugged at the heart of his fellow soldiers. In his eyes, in the sound of his voice, in the way he carried himself during the interview, the desperation known and experienced by everyday citizens in Germany became more real and overwhelmingly sad. In this translator the American soldiers felt they were able to see the fear, anger, and hatred felt by so many German citizens who left Germany to save their lives.

While the translator's story was not the only one like that coming out of Germany, the men of the 175th never personally saw innocent civilians being mistreated or killed by the Germans. It was bad enough to hear stories like that of the translator. This is not to say that the troops saw no evidence of the mistreatment of civilians who were tortured and even eradicated by the Nazi Party in power. The closest that they got to finding empirical evidence of the horror the German authorities were able to inflict on their own civilians came as the end of the war neared.

Since they were front-line troops, they participated in the capture and liberation of some of the first concentration camps. Another of Birch's buddies was Private Payne who was Catholic and had invited Birch to go with him to attend Catholic Church services on one occasion. When their unit entered one of the concentration camps in the area around Celle, Germany they placed Payne and Birch on duty together over the camp. Their unit just happened to capture it and they were planning to move on the following day. Before they could continue onward though, they needed to leave enough troops in place until the Grave Registration Services troops could come along behind and take care of the dead and all of the associated issues and problems that came to light through the capture of the camp. From all appearances, this camp was constructed primarily to imprison Polish citizens behind its enclosures.

The conditions the Americans witnessed at the camp were far worse than Birch thought he would ever see during his tour of duty. The conditions under which the people were forced to live during their sentences as slave laborers brought to Birch's mind the living conditions much like those of a hog pen. He was overwhelmed by the odors and wished at the time that he had access to a gas mask just to move around the camp compound. It was unbelievable to him and the others that anyone would still be alive in that place, yet there were people living, or rather existing, still in the facility. As he looked at the living conditions, he thought to himself "These people are still living here, how is that possible? They have been forced to live here and turned into people who have lost all hope of life. But with the loss of hope comes the loss of the ability to care how they live, or perhaps even care whether they live or not." To Birch and the other men of the 175th it appeared that those imprisoned in this place had had the basic necessities of life withheld from them for a long time. Birch thought about how horrible and unbelievable it was that one group of people could do such things to another group, or other groups, of people. No wonder the translator wanted to kill the sniper he was assigned to interrogate. Birch wondered what happened to the translator's family that caused him to be filled with such hatred. This was the foulest smelling, horrible place he and the other men of the 175th ever witnessed to this point.

As their battalion finally started to move forward, making its way further toward the east, they came across an abandoned hospital which looked as though it had recently been one of the cleanest, neatest hospitals one would ever want to enter. From their inspection and subsequent conversations with other troops entering the area, they learned that this was one of the hospitals used by Germany for the purpose of raising their perfect "racially pure" people. The practice the Germans followed for doing this was to select women and men of high breeding, "pure bred" men and women. The hospital had rooms, or what some of the men preferred to call cell blocks, on the bottom floor that were used for temporarily housing the women initially chosen to be responsible for raising the babies. What the 175th Regiment had stumbled onto with the discovery of this hospital was a part of the larger nationwide *Lebensborn* program that was first introduced in Germany in the mid-1930s. The word *Lebensborn*, which can be translated as "fount of life," was a program which encouraged "racially fit" women to bear children for the Third Reich. The program also protected infants who were believed to model the Aryan ideals publicly elevated by Nazi Germany. The *Lebensborn* program, which ran from 1935 through 1945, remained secretive and included the provision of secret birthing facilities, such as the hospital that the men of the 175th just found. In addition, the program provided hidden identities for the people involved in the breeding program. Most horrific of all was that the program included the kidnapping of hundreds of thousands of children. The men of the 175th had heard of the Germans' efforts at breeding a racially pure population, but they really had no idea of the details of the *Lebensborn* program, how widespread it had grown to be, or the depths of debauchery to which it would go just to achieve its ends.

The program had roots that reached all the way back to the First World War. One result of that war was that the German male population had been pretty-well decimated, which helped to bring about a sharp decline in birth rates across Germany. This was considered by the Nazi Party to be a national problem. When Hitler and the Nazis came into power in 1933, they brought with them plans for establishing a new world order. This new order was to be one in which Nordic and Germanic Aryans would reign supreme, and rightfully so in the Nazi's eyes, since they were regarded by the Nazis as the superior races on Earth. To carry out this vision of making a totally Aryan Europe, those in power believed the first thing they needed to address was Germany's genetic shortage. Heinrich Himmler, the head of the infamous German SS (*Shutzstaffel*, meaning "protection echelon"), was convinced that the main reason for the decline in the birthrate in Germany was abortion. Therefore, in 1935 he decided to counteract this trend by making abortions of children who were considered "racially pure" a less appealing option. The way he accomplished this was to provide alternatives to the mothers of such children. In the cases of those women who had the means to prove that their unborn child met the racial purity standards of the Nazis, they were provided resources to give birth to their children in secret, comfortable facilities. There was a condition attached to this, however, which was that the women were compelled to do this under a contracted agreement whereby their babies had to be given over to the SS once they were born. The SS then assumed the responsibility to educate the children, indoctrinate them in Nazi ideology, and then give them to elite German families to raise.[2]

At first German military were urged to have children with Aryan women whether they were married or not, but with the progression of the war this became mandatory. As casualties continued to adversely affect the German male population, SS officers were given orders to marry and reproduce. The Nazis even encouraged women in occupied countries to have children with German soldiers. As the Third Reich moved eastward, the *Lebensborn* program grew to include wholesale kidnapping. Any children who happened to be regarded as racially pure were taken from their homes and families, and placed temporarily in *Lebensborn* homes. Subsequently these children could be adopted out to German families. The *Lebensborn* program eventually involved dozens of birthing centers throughout Germany and occupied countries. Those overseeing the program even advertised these homes as places where unwed mothers could escape social ostracism, ensuring a bright future for their children. In occupied territories, they were also promoted as places of refuge for women to avoid locals who resented their own treatment by the Germans and in turn resented the special privileges given to women who were pregnant with the "racially pure" children. Just by seeing the hospital that they happened upon was enough to cause a quiet, appalling shock to fall over the men of the 175th as they quietly worked their way through the remains.

The soldiers did not really have enough time to process their feelings about what they saw and learned at the hospital when, not too much farther on east from the hospital, their unit came to a hill on the top of which sat a large barn along the side of the road. The unit made its way up the hill and immediately sensed something was very wrong and had been so for some time. When they approached the barn, they all noticed an awful stench emanating from within, betraying something sinister and dark meant to be kept hidden and sealed away from the public. They could see the barn's doors had been closed and locked and the building had been burned out. The barn, like so many barns in Germany had walls built of stone, while the roof was made of timber and other materials. It was into such a barn that the Americans needed to enter for the purpose of investigating what they feared were the remains of a horrific act. They stopped their convoy and the commander got off, ordering the men to set up machine guns in the event enemy soldiers were somewhere near in the area.

As they entered through the barn door opening to examine what was causing the stench within, they were shocked and speechless to discover that inside was what appeared to be hundreds of charred human bodies. Their corpses had been stacked on each other up higher than the height of the barn doors. The stench was overwhelming, sickening, and caused more than one of the soldiers to retch. The sight was unbelievable, and it seemed impossible that anyone could get that many people into a single barn of that size. When the men of the 175th looked inside it was appeared as though someone took the people and threw them in like corded wood. From what the soldiers learned from the people living in the nearby town of Gardelegen, the Germans had crowded the people into the barn in a standing position. Then, when they had managed to fill the barn with the people, they set up tanks around it and set it on fire. When anyone tried to break out through the doors, or any other section of the barn, the Germans fired on them with machine guns. The result was that bodies were seen piled up and sticking out everywhere. The men of the 175th, along with the other units who had joined them on their trek east, learned that the people in the barn were Jews.

What they all saw and smelled was unspeakable in its reach. It was one thing to hear of atrocities on the part of the Nazis, but to stand in the very place where such depths of evil were perpetrated against a people, simply because of who they were, was unbelievable. It felt wrong even to stand there on the very ground where the people lay, stepping over and around them. As he did so, Birch recalled having to walk over and step around the bodies of his fallen brothers on the beach at Omaha back in June of 1944. It would be unspeakable to treat a criminal in such a fashion, or even to treat an animal in this manner. Yet here they were on top of the hill walking among the remains of a barn beside the road, where hundreds of people spent their last moments together burning alive while an enemy without prepared to shoot them if they made any effort to stay alive. One's sense of preservation would necessarily have driven any person to try to get out and escape the inferno. This excelled all the things that made Birch sick in the depths of his being. Just as it was beginning to seem as though he might make it to the war's end and his part in it, he and the others were thrown into viewing this literal hell on Earth brought about by nothing more than pure, unadulterated hate. He thought of how no creature on Earth, but humanity could perpetrate such a horrendous deed as this. Animals kill to survive, to eat. Humans are the only ones to go so far as to kill out of a pure, deep-seated hate. It was a matter of humans using tortuous modes of murder for the sole purpose of the extermination of other humans. It seemed too horrible to think about, let alone to speak about. He thought to himself "How could anyone ever begin to grasp what I have just witnessed? How could I ever begin to communicate such an atrocity to people back home?"

The division's commanding officer called ahead and informed his superiors that his unit was going to be delayed for a while. He explained the situation, and upon receiving orders, sent a group of his soldiers back down the hill to the town. He sent them with the following order, "You get every man, boy that is able to walk and bring them up here!" After some time and effort was taken to bring the people up to the barn the major gathered them together, glared at them, visibly livid, and shouted, "People like you that would let something like this go on in your community, you ought to be treated the same way! You are going to drag everybody out of there." Then he called for a Graves Registration Services (GRS) crew to come up to the location while the men of the 175th stayed around until the GRS crew arrived and set up equipment. The GRS troops brought bulldozers, dug trenches, prepared graves, and then made the people of the local town drag the corpses out of the barn and place them into sheets and ready them for burial. Then they took care to wrap the bodies in sheets and place them into the graves they had prepared for each body. By the time the men of the 175th left there they had been informed that over 1100 bodies were laid to rest in the graves. All the graves were filled. Even then, at that point in time, anyone who went up on top of the hill to look had to hold her or his nose. When he recovered from the shock and horror of what he saw Birch was amazed at the organization that pulled together all the resources to dig the trenches and assemble the materials from wherever they were being stored. Even more amazing was that they seemed to have ways of getting the necessary resources together in short order. It was enough to enable him to begin to appreciate the difficult, yet significant work performed by the GRS troops.

Unknown to the men of the units helping to clean up the site of the burned-out barn and burying the dead found there, this site was part of an effort on the part of the Germans to hide their atrocities from the rest of the world. It all began earlier when the Germans became aware of the U.S. Army's crossing of the Rhine River as part of its move into central Germany. Wanting to destroy evidence of their atrocities, the German SS ordered that the camps associated with Dora-Mittelbau be evacuated in early April 1945. While they intended to move the prisoners by train, or by foot if needed, to the camp at Bergen-Belsen, they had to stop near the town of Gardelegen and unload the trains, because the railways were too damaged by the Allied air raids for them to continue running trains. Since the number of prisoners from the camps numbered around 4000, they greatly outnumbered the German guards. This led to the Germans making the effort to recruit civilians from the local towns to help with the management of the prisoners. This is how the men and women from the towns' fire departments, Luftwaffe, older home guards, Hitler youth, as well as other civilian organizations became involved with guarding the prisoners as they tried to march them on to Bergen-Belsen.

On April 13 over 1000 of the prisoners who were too old and

weak, sick, or injured were unable to continue any further. Therefore, the German military and civilians took them to a barn outside of the nearest town of Gardelegen, marched the prisoners up the hill and forced all of them into the barn. After cramming everyone inside, the Germans set the barn on fire using straw soaked with gasoline. As the flames became unbearable people naturally sought to escape, even by way of digging under the walls of the barn. Each one that made it through was killed by the guards.[3] The following day the Germans, both military and civilians, returned so that they could destroy evidence of their actions. Their intention was to incinerate the corpses remaining in the barn and kill anyone who may have been left alive. However, the U.S. Army had been moving quickly and arrived in the area too rapidly for the Germans to have time to destroy the evidence. When the very first U.S. soldiers to arrive approached the barn it was still smoldering. This was the site that the men of the 175th and other Army units witnessed outside the town of Gardelegen. Nothing that they had seen to date even began to approach the depths of evil and terror that they saw here. War itself was bad enough, but here was a group of people, Jews, who were imprisoned and tortured by another group of people, Germans, simply because of who they were. Moreover, the German military involved German civilians in their atrocities, and made them serve as "auxiliary forces" to help get rid of a people they came to despise. It was enough to make one sick and in fact many of the Americans who witnessed the site, and the clean-up operations, became physically ill.

Before coming to France and Germany Birch believed that he could never intentionally kill somebody. That seemed unimaginable to him. After he had been there for just a little while all of that changed. Furthermore, it did not take very long for it to change. From the first day at Normandy, he felt certain that he could have walked up and shot a German, and it would not have bothered him at all. In fact, he found that he "itched" to kill a German, so much so that one day he stood over the shield of his anti-tank gun and watched as some German soldiers crossed the roadway a little distance down in front of him and his crew. The German soldiers were trying to get around and outflank the Americans. As he watched them, Birch took his rifle, stood there, and laid in on the gun shield for stability. Placing his rifle on the shield he took careful aim and did his best to pick one or more of the Germans off as each one crossed the road in haste. It was a matter of timing since the German soldiers knew the American troops were just down the road, and so they tried to stagger the timing of each one's crossing. Though he tried his level best, Birch did not hit one. Now walking off the hill leaving behind the barn that stank so badly, he was shaking, not from fear or hunger, but from hatred of an enemy that could be so depraved at its very core. Had this been the day that he fired on the tunnel and bunker that contained the large number of Germans and killed so many of them, he would probably have walked over to look at how many he killed. On this day it would not have bothered him at all to ogle the bodies of dead German soldiers.

10. Reflections and Musings

THE FURTHEST EAST my dad, Forest S. Birch, and his unit traveled was to the Klötze Forest, which the 175th was charged with clearing out. Klötze was the general area where the concentration camp, Lebensborn facility, and barn were located, though it was not possible for me to determine the specific location or name of the nearest town for each one of these facilities. About the same time that they reached the Klötze Forest, my dad and another soldier were promoted to sergeant in the field. The unit commander had one set of stripes to give Dad, whose promotion was dated first, but the other man was going to be coming back to the United States before Dad. Believing there was enough time, Dad allowed them to give his stripes to his fellow soldier and planned to receive his own sergeant stripes before the time came to leave for home. On the battlefield where conditions changed almost constantly, very often things did not turn out as planned. This was the case with my dad's plan to wait on his stripes. It turned out that Dad was scheduled to come back to the States before his stripes arrived. Since this was done on the field of battle, the records were not organized well enough to assure the stripes would follow Dad home, or that the promotion would show up in the unit's official records. While technically he was a sergeant, the records that I have been able to research list Dad as a corporal. Tracking the promotion was complicated by the fact that the 175th remained in Europe until January of 1946 when it was demobilized. Therefore, Dad returned home with an armored division, which was not actually his assigned unit during his tour of duty.

The 29th Infantry Division remained intact for a short time, but was deactivated at Camp Kilmer, New Jersey in January 1946. Eventually the division resumed its status as a National Guard unit. The men of the 29th proved themselves to be every bit as capable as any of the U.S. Army units on the field of battle. The fact that the unit had been a National Guard unit in no way detracted from their performance, but this part of its history was originally of genuine concern to the division's commanding general, General Charles H. Gerhardt. In fact, Gerhardt's approach to getting the job in the field done was to be aggressive. This may be the very reason why the men in this division and its associated regiments always seemed to get assignments placing them in a position where the odds were stacked against them. In the end Gerhardt was overwhelmingly convinced that the men of this former National Guard division were more than capable.

After watching a 50th anniversary special on D-Day, Dad was asked how he felt about D-Day and the war in Europe. His response, "What do you mean how do I feel? Of course, what I saw today brings back a lot of memories and it's sad that you have to have such things as that go on in this world. It's sad that we have such 'heathenistic' I guess you would call it, people that want to create something like that. The sad part of it is so many American soldiers died for freedom where you had a maniac over there taking over all the countries. And you see how these people appreciated it...when you would go in and take a town how they would appreciate you, how they really showed their appreciation, their love. I mean you felt like a giant. Of course, the next battle was another thing. You grow up...you grow from a kid to a man overnight!"

One thing I noticed over the years when talking to Dad about his experiences in the war is the irony to be found in how things during wartime often worked out. One example had to do with Dad's encounter with the German Tiger tank, which I recount in Chapter 7, "A Tiger Near Aldenhoven." Throughout his recovery period, Dad assumed he would be sent home due to the nature of his injuries. Instead, he went back to his unit five weeks to the day after the German tank shot him. Somewhere in a train station, as Dad headed back to his unit, he ran into one of his cousins who had been badly injured while cleaning his gun. His cousin's gunshot wound ended up sending him home to the States while Dad's wound from the German tank artillery allowed him to go right back, almost to the exact same spot where he had been wounded. Maybe the ironies during wartime are more pronounced precisely because they occur during wartime. Perhaps the ironies most of us experience every day do not typically seem to be game changers in our lives, while the ironies in wartime often become turning points in peoples' lives. This happened to Dad just because he switched foxholes with another soldier who kept pestering him to do so. Dad did it to get the guy off his back, but it saved Dad's life.

I have often thought about what it must have been like for Dad when he went to fight in World War II. He was a young man, just twenty years old, when he entered the service and began basic training. He turned twenty-one during his time living and training in England and then turned twenty-two a little over a month after his landing on Omaha Beach in June of 1944. When I was that age, I was finishing up college and pondering going on to graduate school at my Alma Mater, Miami University in Oxford, Ohio. I could not imagine experiencing the things Dad did at the same age. Just to hear about his experiences during his tenure in Europe amazed me many times over, especially given that he faced at least three situations which, by virtue of their severity, should have caused certain death. Dad always maintained he never really knew anything about life at the time he entered the armed services. Still, he was put in charge of ten men almost immediately due to the rolling loss of life in his unit. Having this thrust upon him, he functioned and relied on his training. Even then he walked on the line between life and death several times during his time in England, France, and Germany. His sole focus was to get out of the whole ordeal alive.

Normal daily routine is another thing about which I pondered, and I cannot imagine living without it. I realize Dad and the other soldiers had daily routines which were much different than what is normal routine during peacetime back home. The routine consisted of searching for the enemy, trusting very few people, taking time to build trust with those brought into their group anew, and fighting the enemy. These daily activities are not what I would regard as "normal," which is why I believe Dad admitted to being depressed several times during the war. The most depressed I ever saw him during my years growing up in his household was on two separate occasions. One was when he lost his own dad, Levi J. Birch (I always called him Poppy) to death. The second time was when my dad's younger brother Mac, who I mentioned in the first chapter, fell off a bridge into an overflowing creek at night. Every day after work, I saw my dad go to the creek to help in the efforts to search for Mac, who was found some days later. He had drowned. Although Dad was not a man prone to depression, when he recounted his experiences over in Europe during World War II, he made mention more than once of being depressed and afraid. This makes me think that his experiences were much worse and had a more pronounced effect on him as a young man than he ever revealed. This is the dark shadow behind the accounting of the experiences.

What is it that makes some men, like most of the soldiers with whom Dad fought, set thoughts of self aside and function according to their training and preparation, while others, like the two young disobedient soldiers, abandoned their position and refused to return to their post when Dad, their group leader, gave them a direct command? How was it that the right mix of talents came together on Hill 108, Purple Heart Hill, such that a hunting guide was able to use his acquired talents, slip out, and slit the throats of Germans guarding their exit? The hunting guide did not receive these skills in basic training, but rather brought his skills with him into the war to use when the situation demanded it. He combined his previous skills with those he learned in basic training. The same might be said of the young American soldier who had been born in Germany, but through a series of events ended up in a U.S. Army intelligence unit serving as a translator. The loss of his parents in Germany was a tragedy that in the end enabled him to face situations and bring something to them that could not be taught, which was his own deep-seated animosity for the German military and the Nazi party. It was more the case that the animosity was ingrained in him by way of the actions of the Nazis. While he was an excellent translator by virtue of his being born and raised in Germany, his anger which was freely manifested to the German sniper that my dad's unit caught, certainly seemed to play a role in causing the soldier to answer the Americans' questions without hesitation. Often, I have wondered if the young translator routinely demonstrated his animus to the Germans who were brought to him for interrogation or whether there was something different about the sniper that they captured and brought to the translator that day.

In some of his reflections on the day we filmed his story Dad told of how he and his men tried to sleep in barns, basements, or any possible shelter they could find when they were near a village. Wherever their temporary quarters happened to be there were many nights when they tried to joke and talk with each other, just to keep one another's morale up. Still Dad acknowledged that during such times he would wonder "Well why? Tomorrow I may not even be here." Furthermore, he was sure this same question repeatedly came to the minds of all the other men. At such times joking and talking light-heartedly about anything and everything, or perhaps even nothing specific, seemed to be antithetical to the things that they were witnessing on the ground every day in Europe. Dad would ask himself why, but at the same time he realized that he had to joke, to chat about what his everyday life had been like before he came to fight in Europe, if only to assure himself that life could still have a measure of normalcy, even amid something as horrific as World War II. Were he not able to do this, he likely could not feel any semblance of everyday life. The memories and stories somehow provided a sense of normalcy and routine so important to daily living. Life needed to manifest itself to be the journey of ups and downs that people everywhere experience in their daily routines. Without the jesting, bantering, and light-hearted talking life for Dad and others in his unit would have seemed to become nothing more than a continual downward journey.

On one night Dad and the men of his unit were in an old, damp basement with their guard posted on each door to stand watch. Every so often they would rotate the guard, while the rest of the men sat around down in the basement. Dad recalled a man by the last name of Ward who continually joked around with the others. He was from East Liverpool, Ohio and would sit down with the younger boys who were sitting around, many of them looking sad. Then Boots Ward, as they called him, would get over in front of the guys and pretend like he was playing a horn all the time. He would sit there and act like he was playing a trombone - doo doo doo doo doo. He would then say, "When I get back home, you're going to see in lights Boots Ward and his All-Girl Band." Boots would get everybody laughing with his remarks like that, even if they seemed silly. Things like that went on all the time. It was their reminder of a life with normal daily routine.

Corcoran sat with them and told the men what he was going to do when he got back to "New Yok." He spoke with that characteristic New York accent. Even Sergeant Fardie joined in and told them what he was going to do when he got back to Boston. Routinely the troops who were coming into their units were much like others entering the European theater, essentially kids sent over from the United States to join others already on the field of battle fighting a world threat. The reality of it was they were replacing others who had already given their lives or their well-being in this struggle for the sake of freedom. These younger men were typically around eighteen years old, away from home, and new to the battlefield. There they were, sitting around together scared to death, and Dad sat there with them. As he watched those men, he realized that here he was, responsible for these young, scared kids. Everybody watched after everybody. Dad once said about his experience preparing for his mission, and after landing on Omaha Beach, "You grew up like that (snapping his fingers). You would go in as a kid and overnight you would become a man. If you didn't, you didn't last very long." Years later, on the fiftieth anniversary of D-Day, one of Dad's granddaughters asked him how he felt about killing when he was a young man headed to war. His answer: "Now you say that you couldn't kill somebody! I didn't think I could kill somebody, but after I was there for a little bit and that didn't take very long. From the first day at Normandy, I could have walked up and shot a German, and it wouldn't have bothered me one bit. Not one bit. I itched to kill a German."

Though I had to surmise what Dad must have thought about his hopes and dreams while fighting in Europe, I am certain from knowing him and what he said of the war that he thought about these things. Certainly, one of these would have been the desire to marry and start a family of his own. The year after the end of the war and its celebrations in 1945 my dad married my mother in 1946 on her birthday. It was a bit of irony that my mother's birthday is December 7th. She had been celebrating her birthday as a young girl fourteen years of age when the news came later in the day that Pearl Harbor had been attacked and that many American lives had been lost. Later, at sixteen years of age, she had to go live with a minister and his family in Arizona because she had developed bronchiectasis through having pneumonia nine times as a child and the doctor said she needed a warm arid climate. At the age of nineteen she and dad wanted to marry. Her doctor told her that even with the bronchiectasis, she was strong enough to bear children but would never live long enough to raise them. She did and even saw two of her great grandchildren born. On mom's gravestone are three dates, her birthdate, date of marriage, date of death. Each one was December 7. She was born December 7, 1927, married on December 7, 1946, and passed from this life to the next on December 7, 2009. Together Mom and Dad were a perfect fit and worked together to pass on to my sister and me the important lessons about maintaining core values that would serve us throughout our own lives.

One Christmas Eve after I was grown and had started my own family, we all went up to Mom's and Dad's house to celebrate Christmas. There on the thick wood mantle over the stone fireplace were little felt stockings hung with great diligence, alternating red and green all the way across the mantle, one for every member of the family gathered there. In each stocking were two sticks of hard candy. Dad's first order of business that night was to sit us all down and tell us the story of his Christmases at home when he was just a teenager himself. This is the story that I relate in Chapter 1, "A Metronome at Sea." Mom and Dad were both very careful to teach us to be thankful for what we did have rather than to gripe and complain about what we did not have. They were both also careful to teach us to be a person of your word. They taught my sister and me that one must work hard to do well and that one should always give the best effort. Mom and Dad also worked diligently to make sure that my sister and I were better off than they were at our age. While neither of my parents attended college, my sister and I were privileged to attend and graduate from college, and I have been blessed to have the opportunity to attend and graduate from graduate school. My point is that whenever Dad may have thought about a potential future family, he was true to the sentiments that I expressed in the first chapter where I relate the story of his voyage to England on the RMS *Queen Mary*.

Mom and Dad each had a way of making sacrifices for us in ways so subtle we did not notice at the time that they were sacrifices. As I thought back on some of these later, I was ashamed I took what they did on my behalf for granted. It was as though I had become so accustomed to the gentle sacrificial way that they lived their lives for my sister and me, that I did not notice what it cost them. That speaks volumes for them, as they did this as a way of life, and one which they approached with contentment rather than complaint. This has something, I believe, to do with why theirs' has been referred to as the greatest generation. My parents' generation was born within a decade of the end of World War I, when the world was still feeling the ravages of the first war involving nations around the world. Then within a decade of their birth they saw the Great Depression take a stranglehold on our nation. They matured quickly and grew up learning to live in this era of want and need, learning to appreciate the value of a job and of being able-bodied enough to provide, or help provide, for one's family. They learned not to waste anything and made use of everything possible (Growing up it was normal to see tin foil neatly folded and placed in a drawer by my mother for reuse, and I still have bins of old bolts, screws, and electrical fittings that Dad kept after replacing them on the job). Then within a couple of decades of their arrival on this planet they witnessed World War II, which many believed would be the war to end potential wars to come. They used the lessons they learned up to this point to make it through the war, while maintaining their dreams of living a better life free of war where they could establish their own homes and families.

It seems like a sad bit of irony that while my parents' generation successfully worked to make life better for their children, the children of their generation did not always adhere to the values their parents passed on to them. Maybe in such cases it has to do with the fact that the children were not actually required to live the difficult experiences that permanently ingrained those values into the parents. My generation was taught the right values and principles for living life, but for many of my generation these lessons did not stay with them as might have been the case had they personally lived through the things that my parents and their peers experienced. In other words, what separates the greatest generation from the rest of us who have followed is that they lived through the life-changing events that could have wiped them out, and for many people those events did just that. For those who came through those events still intact they were somehow changed by living though them. They became different people, better people, people who had weathered the most severe storms known to humanity. They came through with lessons learned along with buried emotions that tended to work their way to the surface like rocks slowly working their way up from deep underground by geological processes. They were faithful to pass the important lessons on to their children and their children's children. They have been important lessons and many of the children of that generation have taken them to heart, but not in the same way as the greatest generation. This is another way in which I and others of my own generation have taken our parents for granted.

An example of this can be seen in something I asked my dad to do years ago. Dad grew up in a time when the impression was that the movie house was usually a place to behave in ways you would not want to be made known. He saw this from experience and so as he grew up, he avoided movie theaters altogether. The year the movie *Saving Private Ryan* came out I asked my dad if he would go with me to see it, mainly because I kept hearing that the battle scenes were so realistic that several veterans had spoken out that it gave the audience a real feel for what things were like. Dad consented just this one time to go. After the movie he told me a couple of things. For one, he said that the men with whom he served did not curse every other breath as the movie had the characters doing. This is because they were kids too scared to curse in that manner. To be sure there was cursing, and there were those, like Patton, who had the reputation for crude language. However, while the movie had all the soldiers cursing constantly, Dad said that was not realistic. The other thing he said was that the sky was black with aircraft and barrage balloons, as the Allies brought in a great deal of aircraft as well as fleet vessels of the sea. Then Dad told me one more thing that made me sorry I ever thought to ask him to go assess the movie for its ability to simulate reality. What he told me was that the movie had dredged up emotions he had not felt in many years and that he did not want to feel.

He did not elaborate but tied to those emotions were memories that he was forced to relive. I was so selfish in wanting to know what it was like on Omaha Beach that it never occurred to me that this might not be a good movie for my dad to view. Could he tell me whether the scenes were realistic? Absolutely, because he lived it as one who landed at Omaha and made his way through France and Germany facing a ferocious enemy. Was it worth it to have him assess the realism of the movie for me? Not at all. I would much rather have gone on wondering about whether the movie realistically portrayed the war than to have my dad relive awful experiences associated with the emotions stirred up by the film. For years I have regretted taking Dad for granted in that I again wanted him to do something for me, which was not so important or necessary.

In telling Dad's story I did not say anything about how God watched over him from the time he left for England until he arrived back on U.S. soil. There is a reason for this, which is that while he was serving in the war my dad had not yet come to personally know God in such a way as to have a real intimacy with him. As a youth Dad attended church on occasion, but not routinely or all the time and when he went, he went on his own or with friends. Shortly after I was born my mom made a deal with my dad after he had asked her to do him a certain favor. Whatever it was that he had asked of mom she agreed to do on the condition that he would go to church with her. He agreed and therefore got his favor. On the day when he was called on to pay the favor back, he did not want to go to church. One certainty was that Dad very much loved my mother and so he went to church, but he went to church mad. That led to his committing his life to Jesus Christ along with my mother. At that time in their life, I was only an infant, while my sister was just a young child. Accordingly, I never knew anything other than being raised in a Christian home by Christian parents. My sister does recall how things were different before my dad was converted, which included a great deal of drinking on his part. Sometimes I have wondered during those years of drinking whether Dad ever thought back to how drinking had adversely affected his own dad, and in turn how it impacted Dad's life as a result. In thinking and writing about this I have wondered whether, when Dad started drinking himself, he ever thought back to those days when he used to hide in the corn field from his own dad.

My reason for relating this is that if he were here now to look back on the events during that time Dad would without question say that God watched over him and he has said as much in the past. Many soldiers during that time prayed, whether they were very spiritual or not, or had a personal relationship with God through the person of Jesus Christ. Likely Dad offered up prayers like many others did. After all, he often told me that they were scared kids and tended to pray rather than curse at everything. Still, while he was in the service, I doubt that he would have had the depth of knowledge and understanding to know whether God had a guiding hand on him or not. There were at least three times when it looked certain that Dad would be killed and in addition to these, every minute on the front line meant his life, and the lives of his peers, were in constant, danger. Though he was wounded by artillery fire from a German tank, Dad was not crippled from it. He was able to recover and live the rest of his life with a little sample of that shell embedded in his leg. There were other life-threatening events that Dad experienced in the years after he came home from the war. There were three major ones that I can personally recall, since they all happened while I was a youth still living at home. He acknowledged many times that had it not been for his relationship with God through Jesus Christ he would not have survived to see his kids grown with their own families.

A poignant example of how Dad carried the lessons that he learned into his daily life came to me one day when I had a conversation with my wife's doctor. After performing a surgical procedure on my wife, he came out to talk to me about how things went and how long she would need to be in recovery. This doctor had the initiative to work to pay his way through medical school and for a while he worked with my dad at a local paper mill. My dad was an electrician there and this young aspiring doctor worked as his assistant. I knew that and when I mentioned to him who my dad was the doctor smiled, sat back, and began to talk. He told me that he well remembered working with Dad and was very impressed with Dad's approach to work and life. "In fact," he said, "the lessons that I learned from your dad are lessons I have carried with me into my profession as a doctor." He let me know that he still incorporated the principles that Dad taught him into his work each day. My dad had passed away by the time I had the conversation with the doctor and so I was unable to let Dad know how he was still a part of this doctor's life, and indirectly the lives of everyone who this doctor influenced in any way.

The frightened kid, who at one time did not think he could make it through basic training and begged his dad to lie so he could come home, managed to hold steady, do his best, and finish what he started. In the end it can be said of Forest S. Birch that he not only finished what he began, but that he finished well.

Acknowledgements

WRITING THIS BOOK would not have been possible without the support and help of many people. First, I thank my beautiful wife, Debbie, for encouraging me over the years to write about my heritage as well as other topics for which I have a passion. She has been my constant advocate and always the first one to which I turn for reading and proofing anything I write. She excels at this on a level beyond me and I am good at proofing. She is astute at zeroing in on edits that can make my writing better. Debbie is a wife who makes me feel better about myself and makes me a better man. Should you come across any errors in this book I accept all blame.

Of course I need to thank my daughter Kara, son-in-law Dustin, and granddaughter Kierstin for their love, and for support of my endeavour to provide a chronological account of Dad's World War II experiences. When in high school Kara wrote a paper about my dad's experiences during World War II and won a scholarship as a result. Though they never met Dad, Kierstin and Dustin have taken an interest in him. When Kierstin mentions Dad, or "Papaw Sue," you would not know she never met him.

My niece, Brynne, deserves thanks for producing the sketches you see scattered throughout the book. Brynne, who is handy at producing drawings and sketches, was not only interested but wanted to take an active role in this project.

In addition, I thank my many extended family members, from Dad's side as well as Mom's side of the family. These have expressed an interest in Dad's experiences in his fight to stay alive and come home, fighting enemy soldiers who were doing much the same thing. Several expressed an interest in reading the book and wanted to know when it would be available.

Finally, I offer a special thanks to Gregory W. Siewny, M.D., Chairman of the Board of Trustees, Atrium Medical Center, for his endorsement of the book. Though he did not know Dad long, he got to know him well. Dr. Siewny spent time working his way through school. Part of that time he worked assisting Dad who was an electrician at a local tissue paper mill.

Appendix I. Family Photos

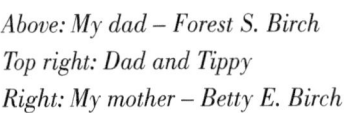

Above: My dad – Forest S. Birch
Top right: Dad and Tippy
Right: My mother – Betty E. Birch

Left: Mom and Dad
Above: Levi J. and Mae Birch

Left: Levi J. and Garnett Birch

157

Right: Dad and Poppy
Below: Homeward Bound

Appendix II.
Transcript of On-Camera Interview with Forest S. Birch

Note: This interview was conducted on June 6, 1994, which was the 50th anniversary of the D-Day invasion. Different family members asked questions. Those questions from our family are shown in bold. I have added some explanations for clarity, which are placed in parentheses. Everything else is told in my dad's own words.

Okay Grandpa, tell us about D-Day.

It was approximately 1:30 p.m. (local time in France on June 7, 1944) and I am sitting right about two miles from the Normandy Beach, out in the English Channel. I was supposed to have gone in on the beach at 9:30 a.m. but was still sitting in the ship waiting. In other words, our instructions were that the 29th infantry division would hit the beach. We had our mine destroyers, our engineers, and all of these troops that goes in ahead of the fighting troops or with the fighting troops and clears a way for equipment to get on shore. They met much higher resistance, because there were more German troops than they thought, and so the plans didn't work out like they thought. They had this worked out in what they called waves; the first wave would be your engineers, your mine destroyers, and your road crews who would make ways for the vehicles to get on shore. That was the main thing, to get some tanks, other vehicles, and different things that were needed. These things did not happen until later in the day, but they were supposed to happen earlier.

We were in the third wave, which all of the main troops were ahead of us; the infantry boys, that actually did the "real" fighting. Our unit was tank fighters. We had a crew of ten men on each gun, and there were three guns in a platoon. And so we were to get on shore and then set up, we were taught to set up our guns in a diamond formation (They would set up three guns in a triangular trap in order to better assure taking out a German tank.). We had communication troops with us that strung phone lines...everywhere you went there were phone lines being strung, where we would keep in contact with the command post. At the same time we had radios, jeep-mounted radio equipment with us at all times. But we never got on shore until around 3:30 p.m. That was Omaha Beach. We were supposed to go cross the beach to a road that was to be prepared for us and go up a hill (this was quite a steep mountain along that beach there), and that climb was almost straight up. And that has been shown on TV; you've seen pictures. And lo and behold that mountain came down like this (into a v-shape at one point). And over here in the middle of it was the beach, from what little I saw, because I wasn't looking at any scenery that day. But the mountain came down like this (to a v-shaped valley). And there was a valley in here. It was rough, but that's where they made their road.

Anyone that heard that hero that spoke on TV today, he was the one. He was the highest decorated man. Because they was all pinned down. They was killing them there on the beach just as everyone hit shore. And he took a few men, eight or ten, and located this low spot. And that's where they made a road to get stuff off of the beach and get it up on the hill, on top. On top there was a little town. By memory I cannot remember the name of the town, but it was a small town; it was not very large at the time. There was very bitter fighting, you know, there was fighting terrible. The equipment was lined up there on the streets bumper-to-bumper, waiting to get through, waiting for the infantry men and the machine gun crews to get in there and clean out everything. It where it was so bad that we had to get out, leave our guns and everything and start fighting individually too to get it cleared out. They did get it cleared it out before dark. When we got it quieted down to where still you could hear sporadic gunfire, but it was safe enough for us to go on through, we had to walk, and watch, and hunch down. We were just kids then you have to remember, we were just scared kids, like anyone would be.

But a lot of these (enemy) machine guns were "manned" by women. The Germans forced some of these women to man the machine guns when they moved out. So they had moved back and set up a heavier defense further back. So our particular group (I can only speak for our group because that is what I was trained for, and what I was told to do) had been instructed to assemble at a certain place. According to the intelligence units there was an apple orchard, a good-sized apple orchard, just outside of this town a little ways. And when we got separated, or lost our crews, or lost our commander, we were supposed to go to this orchard. I had a map, each person had a map on him that showed him how to get to this apple orchard. That was our first destination - from the time that we left the boat we was to reach that apple orchard at all cost. We reached the apple orchard, and never lost a man in our own particular crew, but there were hundreds, even thousands of men lost, and left down on the shore. So that was out first "victory" although we were a bunch of scared kids. From there, then, we regrouped, tried to get our nerves under control, and set out for the next day.

For the next ten days we fought bitterly; it was a bitter fight. It was just kill, kill, kill. We just moved a little bit in a full day's time. We never moved but just a few miles in. By then we were getting so low on troops that we were going to need reinforcements. Of course, they could not get reinforcements in until they could get the beach established so that it was safe to land. But on June the 18th, I think it was the 18th, we were bogged down under heavy, heavy fighting by Nazi troops. They had brought in some of their crack troops, their top trained Nazi troops. They just put us to a "stand still"; I mean we just absolutely could not move. So we were lined up there. We would fight of a day, and when dark came we would lay there and wait for the mortar shells. When they would fire mortar shells you could not hear them. You could hear them go off, but they were silent coming in. You could not hear them to determine where they were going to land. An artillery shell, though, you can hear from the time it leaves the gun to the time it lands. As long as you could hear that, you could tell whether it was going to land pretty close, or whether it was going to go on overhead. You could learn to judge, and we got used to that, so that when we heard artillery fire, we could pretty well tell whether it was going on over, or going to fall short, or hit in the area. Of course, you always took cover. But, when fired, you hear a mortar shell go "pop" over there, but it was silent on its way over here, you could not hear anything in between. Those are the ones that get you; those are the ones I was most fearful of. And then the Germans had on their tanks as well as they had some artillery pieces, had what they called 88s (The Tiger tanks fired the 88s.). They have got one of the most frightful sounds to them, I mean your heart almost fails you when you hear it fire. It is a screaming shell that will absolutely...it will put you out of your mind. They could fire direct with them, what I call direct is that they could sit and shoot straight (like a bullet), or they could angle (like lobbing an explosive shell). It all depended on their distance, but the German 88 was one of the most vicious shells that could be fired at troops. It would almost take your sanity, and they used those things frequently on us. When that thing is coming at you, you can't tell anything about it, because if they shoot directly, if they are shooting direct in (of course all of these are explosive shells; when they hit they explode like a bomb) you cannot ever tell where they are going. The screaming sound they make sounds like it is coming at you every time! And those things really "made you need a psychiatrist".

At night we stayed put. What we learned was that those lit-

tle bitty shovels that we hated so much that they gave us in England...I took all of my training in England. I stayed in England from May of 1943 until June 6, 1944. I had the best of sleeping quarters. I had the best of food that the army produces. I had a recreation area that could not be beat. They had taken over a hotel in this area and put our unit in it. We went down a long set of steps, I never will forget this building, you would go down a real long set of steps. And then you cross the main street, and then you would go down another real long set of steps onto one of the most beautiful beaches you could have. And we had it to ourselves, except at night when the civilians came out and joined us. It was a long beach and this town set right down in a valley. I do not know whether they changed the name of it or not; I cannot find it on any maps. But the name of it was Saint Ives, England. It was not a very big town; it was a fisherman's town. The only factory they had there at all made fishing nets. That was my home for over a year. In fact, I became very attracted to Saint Ives. That is where I took all of my training. We would stay there in the quarters, go down to the beach and take our exercises, play volleyball, and this that and the other.

Then about once a month the Navy would come in and we had to load up our packs, load up everything we had on us, get on the trucks and they would take us up somewhere in England to one of the shipyards. The Navy would come in and we would load these ships and we would go out on the waters like we were leaving England; we would go out there to where you could not see land or anything. We would go out there, circle around for a couple of weeks, train on getting off of the bigger boats onto the smaller boats, climbing those rope ladders that you see on the side of those ships. You put a whole bunch of men on one of those rope ladders and try to climb it. It sways out and in, and is constantly going this way and that, and you are not a Navy man to start with. So we lost men in transferring between boats;

storms would come up to roughen the waters. We lost several men in training. Then we would attack. They had bought up all of the homes and moved the people out in a certain section of England. That represented Normandy Beach. They would start shelling that particular section of beach, and then we would go in and attack it. We did that over and over and over and over, until it was almost a dread to where we got orders to pack up and go out.

The Navy fed us good while we were out, because the Navy had the best food of anybody. We loved to out with the Navy because we got "good eating." We got to doing this over and over to where we wondered "Why is it worth it?," because we knew it heart to heart. But see that's what paid off. That is why so many of us are still alive – because they grilled us and grilled us and drilled us and drilled us until we thought "What is it worth... Why are we doing it?" We knew we would be the invasion forces. But we did not know to what extent until about two weeks before it happened. Because they did not tell us anything until just before the invasion. When we got our work done, we could go into town, and spend some time in town just like anybody else. That is about as far as we could go without a pass, because we lived there in that town. And of an evening I had a pass to go out in the evening.

And one day they called. A call went out to all of the pubs and all the places for all the men to report back to the barracks. That was the first thing...we did not have a chance to tell anybody anything. When we went back to our barracks they told us, "Load up, pack up everything." They had trucks lined up in the back; there was an alley in the back. There wasn't no Army vehicles on the street; you did not know anything was going on. Everything was in the alley in the back. They loaded us up and shipped us out of that town before that town even knew we was gone. But word got out, because some of the boys had girl-

friends and somehow they got word through somebody. And of course these people were back in the alley wondering "What in the world is going on"? We didn't know any more about what was going on than they did. But that was the last we seen of if, I mean they shipped us out and told us nothing until we was out on the moors of England. They had what they called a "tent city" set up out where there wasn't no homes no nothing; it was all desolate land. And there we was put up in tents, and was told we would be briefed for our mission for the invasion. From the time they told us we would be briefed for our mission we were not allowed to speak to a civilian. There was a road that went through there, but if any civilian passed we were not allowed to speak to any civilian. We was under guard by our own soldiers; another outfit had come in and put us under guard. We went to the movie, we went with a guard. And when we went to eat, we went with a guard. As we slept, we slept under guard. We never spoke to a civilian from the time I left that town until... say about June 18, somewhere along there that we got regrouped. And General Patton had made the push for St. Lo. Which today they have a parade in St. Lo in honor of their liberation. But that was the first big city that was taken. And from there General Patton and his "wild tank drivers" made their push through. And they moved so fast from then on that the infantry couldn't even keep up with them, and didn't even try. I mean he just, he just, he was "zoom"; he was headed for Paris and he was gone. And all the guys just marched there and cleaned up these little "hot pockets," that's what they called them, you know, to these marker groups, this, that, and the other.

But we were before this happened, we were pinned down, because our troops we was down, our battalion was down to about 300 men. And we were in a field at..., in northern France (This would have been about 7.2 km north of St. Lo, at Hill 108 where the 1st Battalion held ground from June 17-19, 1944. Dad's unit was surrounded on three sides by the German army and were later relieved in the middle of the night by the 3rd Battalion. For their efforts here the 1st Battalion 175th Regiment won a Presidential Distinguished Unit Citation and the French Croix de Guerre). Their fences, what they call fences, is rock and it was built up with rock and dirt, and its solid; it's sowed in grass. It just looks like a dirt mound really but they are about five or six foot high fences you can't see over them, or most of them. So they are that high, and that's what they had their fences in and they had their gateways in each one. But they don't have fences like we have, they are made up of rock and they are so thick. And at night when we would bed down for, and quit fighting, they would just throw shells at one another, or if they was close... but we would be so close to the enemy that night that we was pinned down; they were on one side of that, what they would call a fence (they were blockades like). They were on one side of it and we were on the other. And what we were doing was rolling grenades over to one another. That's about how close they was; you could hear them jabbering over on the other side. Of course, I guess they could hear us too.

But we bedded down and dug foxholes, well we dug holes that we could lay down in. And one guy would sleep, we buddied up. The major was wounded, we couldn't get him out. And we had radio but we couldn't open up, because every time we would opened radios they would just...artillery shells would just bombard us. And they cut our telephone lines; we couldn't communicate to the back areas to tell them we was in trouble. But anyway, this man went through and got help. And about two o'clock in the morning they brought in fresh troops. And secretly they would come and wake us up and tell us to be quiet, easy up, and ease out. And the next morning of course it was a full company of fresh troops waiting for them.

(With regard to how reinforcements arrived...) The guy that

got help was a guide; he was a hunter's guide in Maine. That's what he did for a living. He took hunters out and packed them down in this country who would go up to Maine to hunt. I guess Maine, or some parts of it, is pretty wild, there is an awful lot of wood area, and I guess its uh…from what he says, and he was a hunter's guide cause people couldn't find their way up there and he would take them out. And that's what he did for a living so he was pretty well experienced and he went to the major and told him…we had no way of getting out see, and we was pinned in there and he told the major, he said "I'm going for help." The major said, "I can't ask you to do that." He said, "I can't stop you, but I can't give the OK, if you go you go on your own." He said, "I'm going." And so, he slipped out. He was a little short guy, I never will forget him, he was kind of red-headed, he was kind of sandy haired and he wore a mustache. But he was a short guy, a little short guy. Anyway, he went back and got help. But when daylight come…He was like an Indian, he took his knife, his knife was razor sharp, put it in his mouth and he took off. And they found I think it was about four or five dead Germans, where he had slit their throats. He would slip up on them, and just cut their throats and go on. And that's the way he got us help. And he was highly decorated for his mission. Then we went on through St. Lo, and I mean, that was the breakthrough.

Everything was…, we had our good days and our bad days. The beaches…, if there ever was what you would say was "hell on earth" that day was. I mean it's something you never can… never will forget. And you wouldn't want to see nobody have to do that, because the skies was just as dark with planes, I mean I don't see how a plane could fly up there without having a collision. But they was falling just as fast, ships was being hit, and it was just unbelievable what was out there that day.

You were in St. Ives and next landed on Normandy, correct?

Well, while we stayed out there (in St. Ives, England) to be briefed on what we was going to do, and then a couple of days before, I can't remember whether it was two or three days before June 6th, we had already loaded; we was on ships. We was already loaded. What they were doing they loaded us and we just, we floated up and down...we just cruised up and down the English coast. Because see you couldn't...all the ships that was in that invasion that day came from many, many different ports of England. Some ships would load over here and some would...and then they all gathered. And there was this particular area that they was going to sweep across. So everybody was just floating up and down the coast waiting for orders. Of course, we knew it was going to be June 6th, I mean well we didn't know at that time, cause really it was supposed to be June 5th. And the storm held it over, that one night. And Eisenhower made it...decided to go on and do it then June 6th. We didn't know the hour...we didn't know exactly, but we knew it was right there. That was our time, cause we were on the ships, we were ready, we had everything, and we were just cruising until everybody got together. And that's when they struck...we knew we was supposed to go in then. When we got the word was that morning, that we would be in the third wave and..., cause everybody was in contact with the main headquarters. And, well there was one wave of ships, I mean one wave was supposed to take the engineers, and all the men that would clear a way for the fighting men to get up on that mountain. Which those men mostly were all left on the beach, I mean they just all was killed. But they had to go there and do the best they could. This was supposed to cleared and be a roadway.

On the ship that I was on (According to the morning reports dad's unit left England on LST471 but when they landed on Omaha Beach they disembarked from LST 262. So at some point they switched LSTs.), it was loaded with tons, and tons, and tons, of trucks. Our truck…take like our truck, was about a ton-and-a-half or two-ton truck. You've seen these Army vehicles. They have a seat down the side for the men to sit on. We had ten men that were with that truck. Plus, they went on shore with fifty-five gallons of gasoline mounted on the side in those pockets of those trucks. Plus, all of our ammunition, them shells are approximately, I'd say eighteen inches long, somewhere along there, and they're very heavy, they are about that big around (indicates three to four inches). And we had, most of our shells were high explosive (HE) shells. In other words, they was designed to go through a tank…there was called, what they was called was APHE – Armor Piercing High Explosive. That a way they was designed to pierce the shell of a tank and explode on the inside. That's what that shell was designed for. And that's the type that we had on there. And when they did that, besides we had, they had a big, a big like a semi truck in the bottom of this boat. I don't know what you called that ship that carried that many men (Dad was referring to the LST. His landing LST was LST 262). It had two decks you know, it had on top, it had three decks, you had uh we had jeeps and everything on top, and then on the second deck they had stuff, and then on the bottom deck it had a, one big semi truck that was loaded with bridging material, with bridging material; it was loaded with a whole load of steel for a bridge. Because that ship when it opened up, it let down a ramp, and then you could go off of there. And see we had orders, see that ship…everything was to go off, and then we was to go off. And of course, that bridging material and everything had to go off because it was on the bottom. But every time they would go off well they would sink down in that sand, see, that was on the beach. They would just sink right down see. Well they, you got that bulldog and all them stuff on them Army vehicles, you couldn't stop them and they would crawl right on out and go on.

And then when it come time for us to go off, us, well the first crew that went off it wasn't our crew, but it was one of the other battalion's crews. And they, the ten men was riding their truck like they was told to do. And they was going straight to the top of a hill, they had a road cleared that was supposed to take them up there. And when that truck went off, these ruts had got so deep there, and they didn't have all the mines cleared. And they was down to the...that mine was so deep that when that truck went off it set that mine off. And that truck, those ten men, and everything disappeared right before our eyes. We was standing there ready to go off, I mean they just disappeared. Just one explosion and there was nothing there. Now you talk about unnerving somebody, I mean we went berserk! All there was, there was a little fire left over here; some of them dropped, there was a couple of them dropped. They was on fire. You couldn't tell what they looked like they was so black and burnt. But they run and grabbed them and took them into the hospital. They had a hospital sitting there on this boat. And they took them right back into that. That's all that was left! And here the commander is standing there, "Get moving, get moving!"... "Move, move!" Because, see they were shooting at us! And so then our commander give us... the rest of us walk. You walked beside of your truck. Don't walk up into the truck. Stay with it, but walk beside it. So the rest of us went on then. Of course, evidently that was the only mine...just happened to be. That bridge truck that sunk that down so far and set that one off; it was just one of those things. And that crew was gone just like that!

(On the ship) Oh yeah...we was standing there on deck, I mean the shells were falling beside of us...we didn't know when we was going to get hit. And we would see ships over here on fire. They'd...some of them shells would hit them ships. And we was wondering, scared to death wondering what was going to

happen. And then they…we couldn't get; we would get all kinds of conflicting information. They would say "Get ready we're going in" and then there would be no movement.

(At the time this was happening) Well I was really twenty-one (years of age), then. I was, the next month I was twenty-two. But…that was, that was the first of the whole thing, now we went, we had other bad times. When I was wounded, and I was on further up that was just right close to the city of Cologne. I stood up on the hill, across the river from Cologne and watched that city (get) destroyed. Our own bombers laid that thing flat as a biscuit. Because that was a high point of resistance. I sat up there because we couldn't get across the river. They had blowed up all the bridges, and it was cold. Man, the snow in Germany gets that deep (indicating about 4 ft.). And every time the engineers would put up a pontoon bridge, they would let them get the thing… that river is a wide river… and they would let them get that thing across, and then they would blow it up. And we sat there for, I forget how many weeks, and couldn't get across that river. But we were sitting there every day and watching our own bombers come over and bomb the city of Cologne.

Just out of there was a little town, I can't think of the name of it but anyway it was a crucial town to be such a small town; it wasn't any bigger than Monroe (Monroe, Ohio) I don't guess, if it was that big. But they fought for that town and we sat outside of it, there wasn't no tank action, so we sat back. When there wasn't no tank action we sat back, out of shooting distance, you know, not out of artillery (range), but out of rifle fire. We would sit back and talk and laugh and wait for orders to move up. Because we were always in contact with the command post, through our sergeants or lieutenants. We never kept a lieutenant half the time because by the time we would get one they would kill him. And so our sergeant was usually our commander most of the time.

We had a sergeant from Maryland, his name was Sergeant Long; I don't know his first name, but he was a Maryland farmer. He was a short guy; he didn't seem to care for anything. And me and him started buddying together in England. And he was our commander up until…because we had two lieutenants get killed one right after the other, and, our original lieutenant got killed shortly after we hit shore. And they sent us one in from, I don't know where they came from, I don't know whether he came from the US, but he had just come from school. And he came in there and was going to make us obey Army rules. We were going to shave every day, and you were supposed to salute, and the sergeant told him, and said "You don't know what you're talking about." He said, "We don't have no rules like that over here." He said, "You're going to make your men salute you, and you're not going to live a week." Because see they took their bars off (rank bars off their shoulders). An officer wasn't supposed to wear any bars. Because that gave the enemy (a sign that) he was an authority, he was a leader, and they wanted to destroy the leaders. And so, what they did was on the helmets; you noticed on the helmets, on the back of the helmet there is a white piece of tape, you have noticed in these D-Day pictures, about that long (about 4-5 inches). The white piece of tape on the back, if it was around this way (horizontally) he was a noncommissioned officer, he was a corporal, or he was a sergeant; if that piece of tape was up this way (vertically) he was a lieutenant, or he was an officer (of some kind) or he was a captain. And so that's how you identified your leaders, by the tape that was on their helmets…no bars, they took off their bars. They didn't wear any bars. But he was going to have us behave like Americans did when we were over here. He was going to have us shave every morning, well we would go for a week over there and not shave. We wore old dirty clothes, and that stupid thing come in there and wanted us to salute, and everything like that. He lasted about ten days. Somebody killed him. So, we just went on along with Sergeant Long. He was the most intelligent person…he would always take me with him of a morning up to the command post. I never would want to go and he would say "Yeah come on Birch, go with me." I would say, "No I don't want to go." And he would say "Yeah, I may have to send you back after the crew. And I would say "No" and he would say, "You don't have to worry, its them bullets, as long as you hear them bullets you don't have to worry, it's the ones you can't hear that will get you." I said, "Yeah, I know that." But the day he got wounded, he was up on the front by himself, I didn't go with him. He went up there to get orders (to see what they were to do next) and he got wounded. And that just about really took the heart out of me. I was so depressed, because I knowed that man since we met in England, we had buddied together completely and he was an excellent leader, we watched after his men. He made sure they were as safe as safe could be in a war. And I actually laid and wondered how I could wound myself and get out of it (the whole situation). Of course, I never did pick up the nerve to do something like that but that's how much I thought of that man.

But the final on that town that we fought (for) that we took in; it was a small town, and they would go in and take the town. It was a bitter fight, man they was hauling wounded out of there by the jeep loads. Just as soon as they would take the town it wouldn't be but about an hour and they would say "Move back, move back"! The Germans would come right back and take it again. I do not know how true this story was, but we were getting information that that one little town changed hands three times the same day, and along up in the evening the US took it again. That time they gave orders for us to move up, and we moved up into this town and they moved our guns and they told us where to put them and we moved out guns out into a field. This village was up and had a real high bank and you would have to jump down and you would just be standing up in this field (above the bank) and you could just see the tops of these houses down there, that's how it was. Down two or three doors form where we set our tank guns there was a first aid post in a house.

We set up our tank guns, and we usually set them about a thousand yards apart. We would set two guns back like this (like the base of a triangle) and then one gun way back. Our strategy was that anywhere tanks would travel that was the best route, and that was the only place tanks could come in on that town, they would come in through that field. And sure enough, that was on Thanksgiving Day, and our sergeant, he was Italian and he was a real good cook, I mean he commanded his kitchen and I mean just perfectly. He said "I'm going to see that my men have hot food on Thanksgiving Day" (Nov. 23, 1944). We were looking forward to that hot meal from John Smatana (morning report #306), that was his name. So, we set out gnus up there; we didn't keep all ten men out on the gun because that was foolish to do so. But what we did, we would dig in, we would dig a foxhole big enough for two men, and two men stayed on the gun. We had a telephone and if we could find cover like I said you could see the roofs of those houses, and what you would do is you would jump over a fence and go down this high bank right into this house. So, we put our men down there in that house, two men on the gun. Then if they sighted a tank, they would communicate; they were in communication with the other two guns, and then they was too call the leader. If I was down there, which I was one gun leader, and so that way the men got a little rest.

So, this tank came in, right directly in front of our gun, and these two boys panicked and left the gun and came down to the house. They didn't even call. And they said, "There's a tank up there." They were two replacements, I mean they hadn't been with the unit all the way; they had come in after. So, I told the gunner to get back up there, and the gunner refused to go. He panicked, and this was our first encounter with a tank. I'll put it that way. I said "Well, you guys can be court-martialed for that" and so I took off. I had a guy from New York whose name was Bill Corcoran and Bill said "I'll go with you." Me and him never did click too good because he was sort of a smart aleck, he would smart off. But anyway he said "I'll go with you Birch," because I had to have an assistant, I had to have someone to load it for me (the antitank gun). So, we went up, we crawled over the fence, crawled down in, our gun was dug in, and of course we could look out over that and there he sat. Man, just a picture, and I could just see myself blowing that first tank up. But then at the same time going through my mind was that I could see two or three machine guns at the same time. When he would turn that turret around, I could see them machine guns. I thought "If he gets a glimpse of this in any way, well then he could just mow us down." So, Bill shoved a shell in there, and what (he) done, when he shoved the shell in he was supposed to tap the gunner, the trigger man on the shoulder. That way it is clear to shoot. Well, I waited, I talked to the other guys cause they was going to get him pointed their way the I was going to shoot him in the rear. That's the easiest way to knock them out. When they fired, he swung around, but he didn't swing all the way. And I was trigger happy and before he swung on around, I thought I could get it (the tank), and I fired because it was real close to me. He was as close to me as from here to that garage (perhaps 250 feet). Of course, those guns make a big flash when they go off. Evidently somebody in that tank got a glimpse of that flash close and he swung back, and when he swung back Bill had thrown another shell in and punch me on the shoulder and of course I wanted to get another shell off, and I got another shot off. But Bill didn't *get his arm back out of the way before he punched me on the shoulder and those guns have a 36-inch recoil. So, when I pulled the trigger it broke his arm. Of course, he couldn't do nothing else, so I told him to get out of there and he jumped the fence and went back.*

The guy opened up then because he could see me. Though he didn't know exactly where I was but he had an idea, because his shell hit close enough that when he fired the concussion from that shell blew me from my gun back about ten or twelve feet, and I landed just back on some dirt (Per the battalion morning reports dad was moved from active duty to the hospital on Saturday, Nov. 25, 1944, two days after Thanksgiving of 1944.). My helmet strap was buckled and it seemed like it just about broke my neck. I hit so hard it just seemed like my throat was sore for a week or two. But then I knew that if I raise up, he is going to machine gun me. But I couldn't lay there, I knew I couldn't lay there, because he was searching, he was just going back and forth with his turret looking for any movement. See we rode in tanks, we trained in tanks. We knew what they could see. They had a slot about that wide (2-2.5 inches) and about that long (5-6 inches), two guys and when they're closed in, when they have their turret closed, then all they can see is what they can see through that little slot. That's what they got to look through. If you can hide you can let a tank come right up to you and you can stand up and walk right along side of it. He'll never know you're there. The only thing is, if he has got some buddies over here (away from the tank), they'll know you're there. But they can't see much, and with the training we had riding in these tanks I knew that he just had that slot to look through. So I had a chance to make one jump over the fence, because there was a fence I had to jump. If I could get over that fence before he could pull that trigger on the machine gun, I had a chance. But, when I made that jump is when I discovered I couldn't walk, and so I went rolling down the slope and some of the boys ran out and got me and took me on down to headquarters. I laid there, and this boy from Middletown (Middletown, Ohio), I can't think of his name, he was there that night. Anyway, they patched me up.

They had the town sealed off again; they couldn't take their wounded out. And I know they patched me up. I wasn't as bad as some of the other boys but...and he said, "We don't have any place to put anybody, we'll have to put you back in this potato bin." So, I laid on this pile of potatoes until, they were going to try to get us out after night. But it was way after midnight, it was real dark, and what they did was that they cleared a way that they could sneak them out. They figured well they could sneak out. So, they loaded us up on jeeps, and then they took white cloths and a guy would get in front of the jeep with that white cloth. They had bombed the road too, and the bomb craters were so deep that they had to work their way around those craters. And these guys took white cloths, white handkerchiefs or whatever and they would get in front of a jeep and they led it out of that town back to where it could go on its own. Now remember your driving with no lights. You don't travel with lights in wartime. And so that's how they led us out to safety, with that white cloth. And then they took us back, it wasn't too far back; they had what was called a field hospital. And that's where they operated on me that night. They took me in there and took the steel out of my knee and my hip. And I don't know, I don't remember anything after that. I don't know how long I was there. I don't think I was there for more than 'til daylight, I don't recall. From there they loaded me into an Army ambulance and took me back to a Belgium hospital. And there, I stayed there, I don't know how long I stayed there either. Because I mean I was under medication, and these Belgium nurses, they couldn't speak English. But you never got any better treatment. So one day they came in; a couple of soldiers came in and they had tags (pointing to his ankle/foot) on us. They didn't tag your bed; they put a tag on your ankle. And they picked up my tag and looked at it and said "Yeah this is him." They loaded me up and put me in an ambulance, and I had no idea where I was going, but I kept, I could hear train whistles. After a while they pulled in to this place and they set there a while and they unloaded me and put me on a train, a hospital train and soon I ended up in Paris. I was in a real fancy hospital, I was on the ninth floor and ...beautiful. American nurses, Americans had taken over that hospital and American nurses were there to take care of us, and American doctors. That was Thanksgiving day, and I stayed there until January 1st, New Year's Day (According to the morning reports, #317, dad was returned to duty on December 30, 1944, just in time to celebrate New Year's with his unit. They were then at Pattern, Germany.)

I was hoping all the time that I would get to go home. On New Year's Day they loaded me on a truck, put me on a train and sent me back up to my unit. I got back to my kitchen area. The kitchen area of your unit always stayed back about five to ten miles back from the main front lines. They took me up to the kitchen area and I was supposed to go on up by jeep. And man they was having a field day. I mean they was shooting like it was...Germany sent every plane they had out that day trying to ...it was making its last stand. And they as out there just like they was having a practice on balloons or something. They was shooting them down just like it was nothing. They was just falling and everybody was laughing "Let me shoot." They was getting out there and just having themselves a ball. Now that the...that was the end of the German Air Force, they destroyed it that day. Then I went back up to my unit and we really never had anything to do from there on cause we were looking forward to going on to Berlin.

Around June 18th (1944) when we were dug in for a week or so (This would have been at Hill 108 where the 1st Battalion held ground, surrounded on three sides by the German army, and then were relieved in the middle of the night by the 3rd Battalion. For their efforts here the 1st Battalion 175th Regiment won a Presidential Distinguished Unit Citation and the French Croix de Guerre.)… We was dug in because we was waiting on fresh troops and they was getting organized. And this one boy told me, he come over to my hole and he said, "Birch trade with me, I can't sleep over there." I said, "Oh you'll be alright," because I had me one (foxhole) dug out, had it in pretty good shape. I had covered most of the hole with some old timbers and dirt and I just had a place where I could slide down into it. Then I could lay down in there and stretch my feet out, I had dug it that long. And we would dig a little hole in the side. If we got a letter from home or something, we could read I and lay it up there, nice "home." But anyway, he would come over there and he would say "Birch, trade holes with me, cause I can't sleep." I just figured he didn't have as much coverage over his hole as I did mine. I was trying to protect against mortar shells. So finally I said, well I'll trade with you tonight, I said "I know you can sleep in that as good as you can (this one)." We was only about ten or twelve feet apart. And a mortar shell before morning landed right square in that hole and blowed him all to pieces.

Did you have a buddy that was shaving and felt that he was going to be killed?

(Dad's account here may date to around the last of July or first of August in 1944, due to the fact that at this time his group met and talked to General Patton. Patton came through France near Vire around this time, which coincides with the time that dad's unit was in the area.) My gunner, we got sur-

*rounded again for a short period of time down there in France.
And we slept in a barn that night. And the French over there
would have these great big winery barrels...great big barrels.
And all of them makes apple cider, and every barn you went
into they would have these great big barrels. Man they had
the best cider you ever tasted. But anyway we slept in this barn
that night and he was shaving next morning and had his helmet down there; that's what we had to use for a wash basin,
with water in it and a little mirror sitting up here. He was from
North Carolina and he said, "Birch I feel like my luck's about
run out." I said, "Oh don't feel like that" I said, "I feel like
that all the time." Just made a little remark, you know. And he
went ahead and shaved. And he didn't no more get done with
shaving than boy here come some of the army troops and the
infantry back through there and the lieutenant come in and
said, "Where are you guys from?" We told him and he said,
"Hook that gun up and get out of here!" He wasn't our officer,
so nobody minded him you know. After while somebody else
said "Get them guns out of here" they said, "the Germans are
coming." They said they're pushing us back. So we seen that
they meant it so we jump on the guns and the truck and everybody got on the truck. And we had certain places to sit. All the
men was to get on, but me and the gunner sat on the end (in
the back of the truck, two rows.). So, we would be the first two
off, last two on. And we went back and started back down the
road from where we had come from the night before and down
there was a crossroad. And the Germans had got in behind, and
set machine guns there at the crossroad, which we didn't know
it. And we started down that road and boy the machine gun
bullets were coming down through there like something else.
See you can see machine gun fire. Every other bullet is a tracer
bullet, looks just like a streak of fire. Man it was coming right
down at us you know and man that trucker he slammed on his*

brakes and everybody went off in the back. And I don't know why, we always went off and fell right down in the ditch. But I was on the outside and I seen a hole in the bank over here, a great big hole. And I jumped off of that truck and headed for that hole, and I just fit right back in it. He got off of the truck and instead of falling down in the ditch like he always did, that's what he was trained to do, he comes around that gun (57 mm antitank Howitzer being towed behind the truck), on the outside on the driver's (side), down in front, ran in front of that truck and the sergeant was there saying "Get down man, get down man." And he took four bullets that we could count right across the chest. He walked right in front of that sergeant.

This sergeant I said his name was sergeant Fardie (William Fardie from Morning Reports #216 was promoted to staff sergeant on 6/24/1944, and #351 was promoted to first sergeant on 4/1/1945). He was from Boston, and you know how them Boston people talk. And then he hollered at me and he said, "Birch get over here" and I said "I'm not coming over there." And he said, "We got to see whether Jimmy's dead or not." I said, "Well you can tell he's dead, he's not moving." And he lay there and said, "Birch get over here" and I said "I ain't coming over there." Well, the machine gun fire and kind of ceased. And we could see soldiers, some of our troops down there, and yet I wasn't about to make no move. In the meantime, all of that happened in just a few minutes. Of course, all of our men were laying just as flat as they can in that ditch. We could see them and we could hear men talking down there, and then one of the men over there said "Birch you better get out of there," and I said "Why" and he said "Look up over your head." And I looked up and our troops, some of our tanks had moved in and were cleaning out things down there and there was a tank that pulled up to the fence line along there and had his gun (turret) circling around right over my head. If he had of fired that gun it would have just burnt me to a crisp! When I seen that, man, zoom, I sailed across the road right in the ditch and crawled down to where sergeant Fardie was. And then me and him went out and we could see soldiers standing out in the road down there and so then we went out and checked him. No, no one will never know that (This is in reference to the question whether he ever came to know why the gunner ran around the gun, down the side of the truck, and across the front of the truck in front of Fardie.).

So, we went on down and there was a deep gully, you went down into…there was a deep place, went down in there. Here come driving a jeep in and General Patton with his driver, and he drove in and looked at us and said, "What outfit you guys from?" We told him, and he said, "Well what are you doing down in here?" We said, "We just lost a man," he said "Lost a man"? I said "Yeah." Or I didn't say it but one of the other men said "Yeah, he's laying out there in the road." He said "Well nobody was shooting when I came in." Well come to find out he had come in a different way. He stood there and talked to us a minute. He was kind of a rough talker and he stood there and talked to us a minute. He said "Well, anyway I hate it you lost a man," you know. So, he got in his jeep and took off.

So, we went back out and took everything off, all of his personal stuff off of him and one dog tag, left the other one on him and put his bayonet on his rifle and stuck his rifle in the ground and put his helmet on it so they would pick him up. We left, we had orders to move out. That's what we would do. If we lost a man, or if we found a man we would take their, you were supposed to take their personal belongings off of them, take one of their dog tags and leave the other one, stick their bayonet on their rifle and stick it in the ground with the butt sticking up and put their helmet on it. Then we had a crew come along behind and pick them up. We would turn that in to the commander, and they was the ones that was supposed to turn it in

to headquarters, and they would notify the relatives. And that's
just a small part of what we went through.

Did you ever make it into Berlin?

No. We got within forty miles. We had our (vehicles) loaded
for Berlin. We wanted to see Berlin; that was our heart's desire.
So, we were going to make it to Berlin, well we got to what
was called the Ruhr River. We was within forty miles of Berlin,
and orders came down for us to stop at the Ruhr River. So, we
stopped and we camped there, I don't know for how many days,
waiting for the Russians to come on. Because they wanted to
take Berlin, they wanted to meet us at the Ruhr. Well, all these
Germans, there was thousands and thousands of them, they
came down and wanted us to take them on over. Naturally they
didn't want to be captured by the Russians. We didn't have any-
thing to do so we got boats and boated them across (the Ruhr).
I mean we boated them day in and day out, until the Russians
came down, we took thousands of them over on our side.

We were part of it (the effort to go into towns). See we would
go in, then if this wasn't a tank area, well then we could just
sit idle. They would put you to cleaning up, because there were
always snipers. I can't remember what town it was, let's see, it
was a big city. They went through this but there was still an
awful lot of snipers in there. And there was intelligence that
informed (us) that there was a German officer still in that town
and they found that they had located him, where he was liv-
ing. And me and sergeant Long was given orders...You see we
was in headquarters company, and we was with intelligence,
and anti-tank, and this, that and the other. It was made up of
different stuff. But we was given the orders to go capture this
man and arrest him. He was a high officer, was supposed to be.
So, we went and went to the house and a lady opened the door.
Of course, she knew what we was there for. And the gentleman

came down the steps and as cleaned up and in his uniform, just like it was his dress uniform. And we didn't treat him any way (that was offensive), we had to treat him under the law as an officer. And we took him and marched out and took him in our jeep. Of course, we kept a gun on him but we didn't arrest him (he was under arrest but they did not bind him in any way), or say anything violent to him. We took him down to intelligence and turned him over to them.

A while before we got to the Ruhr River though there was a town we was in. These towns would be like from here to Monroe, that far apart (The distance dad was referring to here was about 3 or 4 miles.). You could be here in one town and look across an open field and see the next town. Well, we was in this town for a few days. It was so muddy, so rainy; the mud was (so deep) in the streets that you couldn't hardly walk. And the tanks was…half the time they tore up things. I don't know what they was waiting for but anyway we was waiting for something. We had our guns set out along a fence and we was in another, an apple orchard behind this house. Our guns was pointed to this other town. If tanks come through there well then we would be (ready). Every morning two houses up, we took over this house. There was a woman, she had a teenage daughter. Her husband was either killed in the service, or in the service, I mean we didn't know anything about it. They couldn't speak English. We didn't have anything (directly) to do with them. We gave them a section of the house and we took the other section, just for sleeping quarters. And every morning I would go down to check the gun crew, cause we changed guard every two hours. Two would come off and two guys would go on. But every morning I would go down. And shortly after we would go down there we would get shelled. I mean man there would be artillery shells by the…for about an hour they would come in there something terrible. And this old man, he would go out, he had a chicken house out there. And we would go through his house and barn, over there the house and barn and everything is built together in a lot of places. Anyway, I would go through his drive. His barn was on one side, part of his barn, and part of it was on this side with the house. We would go through there out into the orchard and he had a chicken house out there and a couple of other little buildings. But he would just smile (nodding his head) you know. He would go out, and I noticed that…and one of the guys said to me "Every time he goes out we get shelled." So, I told our commanding officer. He said, "Check that chicken house out, and see what's in it." So, I went down one morning and got him and took him out there and we went through it, couldn't find a thing. But there was an awful lot of wire, you know, laying around in there. I mean there was all kinds of wire laying around in there. There was nothing there, no kind of equipment. So I let him go and I told the commanding officer. He said, "arrest him, bring him down and interrogate him." So, the next morning I went down and knocked on the door and the old man came to the door. I couldn't speak German and he couldn't speak English so I motioned for him to go with me. I took another guy with me and I said, "Now I am going to take off down the road and he's going to follow me, you stay behind with the gun." So, I motioned for him to come and go with me. He took off, put his cap on and took off with me, you know, following right behind me. I took him to headquarters and turned him over to them. So, we was there, I don't know, two or three more days I guess. But anyway, every morning I would go down. The woman knew I was the one that come and got him, I felt sorry for her really. She would come out there with her apron, she would take her apron wiping her eyes and crying and said, "Papa no come home, papa no come home." And I would just keep walking cause there was nothing I could tell her. "Papa no come home." And then the same…

the next morning would be the same thing. She would wait and you would go down there and "Papa no come home." When we left there papa still hadn't come home. I don't know what they done with him.

What about the American soldier who could speak German, and wanted to kill a captured German soldier?

There was a boy, this boy was a German. He was a German; they killed his parents. And he lived in Germany, I'll put it that way. I don't know if he was Jewish or what he was but they had killed his parents. And somehow through his family he made it to the United States, and he joined the Army. And he joined to get into intelligence as a translator because he was good. I mean he could...he was good at it. But anyway, they caught a sniper, a sniper was in a tree and some of the guys got him and brought him down, and they asked who the translator...who might speak English (German) and he happened to be around. He said, "I can." And man they couldn't (hardly) keep him from cutting that man's throat I mean he was...he didn't want to interrogate him, he wanted to kill him. And they had to hold him back to (keep him from killing him). (They said) "No we want information from him, to know what's around. You're not going to" (kill him). Yeah, he made him talk. That was in the early stage...that was in some of the early stages of it. Yeah boy, he had a knife and that kid, he was...he wouldn't have talked to any of them, he would have knifed every one of them if he had a chance.

Did you see many innocent civilians killed, children and women?

No, none ever. The only things I saw (related to civilians)... we were the ones on our front that captured the first concentration camp. And one of my buddies, Private Payne was his name, his last name was Payne, he was a Catholic. And I know he had me go to Catholic church with him once. But anyway, they put him on duty over at this camp. See we had captured it. I can't remember the name (This may have been Dinslaken, which was a slave labor camp that the 29th liberated. Dinslaken is only 12km from Mehrum, which is where dad's unit was between Apr. 1st and 4th in 1945. Or, perhaps more likely, since they were on their way to the Klötze Forest, it may have been at the camp closest to Celle, Germany which was on the way to Klötze, which was very near Gardelegen.). That's one thing I lost (memory of). It wasn't our duty, we just happened to capture it and we was going to move on the next day. And they had to leave enough troops there until these people…they had a crew come along behind us who would take care of the dead, and anything like that (the camp) that you would capture. And it was full of Polish people mostly. You talk about a hog pen, that was a… I mean you almost had to wear a gas mask to go through it. Stink, and I guess those people had lost all hope of life, and they didn't care how they lived. Of course, they weren't given nothing to live with. But that was the most stinking, most horrible place that I ever (witnessed).

And then on up we captured a hospital. It was one of the most clean pure hospitals you would ever want to walk in. It was one of the hospitals where he was going to raise the perfect German man. They would select women, select men, and I guess of high breed, pure bred. And they had cell blocks in the bottom of this place and they kept them there to raise babies. They would kill the girls and kept the boys.

And then not far from this place, just not too far from this place, and I wish I could remember these names, we was moving. Our unit was moving, we got up about half way to this hill and there was a big barn on the side of the road there. And they had noticed that there was a stink coming from it. Of course, you could see the doors. So, we stopped, we stopped our con-

voy and the commander of course would get off and we set up machine guns there in case there was something around. And there was over 1100 people in that barn, they were dead. They were stacked on each other higher than the barn door. It was unbelievable that they could get that many people in that barn. What they did...see barns over there their buildings was made out of stone and the top part might be timber, but anyway they put them people in there. They crowded them in there standing. Then they set up tanks around it from what was told by the people down in the town. They set up tanks around it and then set it on fire. When those people tried to break out those doors they would machine gun them. There was bodies sticking everywhere. You talk about a mess, that was a mess. They were Jewish people. Our major called ahead and told them we would be delayed a little bit. Anyway, he got orders and he sent a group of our men back down to this town. He said, "You get every man, boy that is able to walk and bring them up here." And he said, "People like you that would let something like this go on in your community, you ought to be treated the same way." But he said, "You are going to drag everybody out of there." And, of course, he called for a crew to come up there and we stayed 'til one crew got there and they set up tanks. And they would bring in bulldozers up and they dug trenches and made those people drag those bodies out of there and of course by that time we was getting them in sheets and things like that. I don't know where all that stuff come from but they had a way of getting it pretty quick. Then they would wrap them in sheets and place them in the trenches. That place was over 1100 I am told when we left there. I know it was full. You had to hold your nose when you would go up there to look. When they opened the door, it was just like they took them and throwed them in.

Now you say that you couldn't kill somebody! I didn't think I could kill somebody, but after I was there for a little bit and that didn't take very long. From the first day at Normandy, I could have walked up and shot a German, and it wouldn't have bothered me one bit. Not one bit. I itched to kill a German. I stood one day over the shield of our gun and they was crossing the road in front of us up there, fixing to try to get around us. And I took a rifle and I stood there and laid in on that shield and done my best, as those men would cross the road, to pick me one and take him out and I never hit a one.

But the reason I got that Bronze Star; it was nothing to be proud of but I was doing my duty. We were setting down waiting for columns to move forward. We were setting there parked along the road under a tree so airplanes couldn't see us. Because our own airplane shot our truck once and blowed it all to pieces, mistook it for an enemy. So, we tried to hide out trucks, and we was laying around there talking, and all of the sudden I got a call to bring my gun up. I wondered why they wanted just one gun up there. We had to go through some rough terrain, up a hillside to get up there. Then you could see across a field, a long field. Well B Company was going to take…they had a bunker over on this hill, there was a tunnel under this hill, you could see it in the open. What they had back in here was equipment and all these soldiers. And they would come out and pin B Company down, they had them pinned down out there in the field. B Company couldn't come back and they couldn't go forward, because there was a little high place out there in the ground. All that was saving them was that little raise in the ground. The field as flat but it went up like that and then lie that (raised out in the middle forming a small mound). So, they (the Germans) couldn't get their guns out there to kill them. As long as B Company laid flat on the ground they were alright, but they couldn't come back or they would get them, and they couldn't go forward. So, they called for me, our gun. I took it up there and there was a fence there and (they said)

"Stick your barrel through that fence." He said, "I want you to fire three rounds of armor piercing." That's a shell that doesn't explode. And he said, "I want you to zero in to where your shell will go right into that opening" (of the tunnel). I could just barely see it through my telescope, and when I zeroed down to get my gun barrel zeroed into that opening, it looked just like I was firing into that high place (in the field). And so, I told him, the sergeant "I'll be firing into my own troops," and the lieutenant said, "You fire!" He said, "You aim for that hole." And I still didn't think my shell would go; I thought it would go right into B Company. But the first shell that I fired, it did miss the high spot and it landed a little bit to the left or right, I don't remember which. And he told me what degree to turn. They was watching with binoculars. He told me how much to give it and fire another one. And I fired another shell. The third shell went right into the hole. Now he said, "I want you to fire three high explosive (HE) shells." He said, "Don't touch it, fire three right into that opening." And we fired three shells into that opening. And when the smoke cleared B Company went over and what was left came out of there holding their hands up. They said I really put a load of them down; I didn't go over there to see. That's why I got the Bronze Star, because I did that out of line of duty; it was beyond my call, because I was a tank fighter.

How do you feel, fifty years later, about D-Day?

What do you mean how do I feel? Of course, what I saw today of course brings back a lot of memories and it's sad that you have to have such things as that go on in this world. It's sad that we have such, "heathenistic" I guess you would call it, people that wants to create something like that. The sad part of it is so many American soldiers died for freedom where you had a maniac over there taking over all the countries over there. And you see how these people appreciated it…when you would go in and take a town how they would appreciate you; how they really showed their appreciation, their love. I mean you felt like a giant. Of course, the next battle was another thing. You grow up…you grow from a kid to a man overnight! Because here I was a young man. I never did know anything about life, really. But yet, I was put in charge of ten men al- most immediately, because of the loss of life. I know we would try to sleep in a basement…I know one night we tried to joke and everything, to keep one another's morale up. And yet you would wonder 'Well, why? Tomorrow I may not even be here.' And I know one night we was in this basement and we put a guard on each door to protect us. We would change the guard every so often. And we was setting down there and it was an old, damp basement, you know. But it was secure. I remember this Carl Ward…I will never forget him because he was always joking. He was from East Liverpool, Ohio. And he would sit there, and these boys would be setting there, and here some of them looking so sad, and they were kids sent over to us from the United States, eighteen years old, away from home, and just new into battle. And they were setting there scared to death. And I would set there and look at those men. Here I was responsible for them. Ole Boots ward, he would get over there and he would pretend like he was playing a horn all the time. He would set there and go (making like playing a trombone) doo doo doo doo doo. He would say 'When I get back home, you're going to see in lights Boots Ward and his All Girl Band'. And he would get everybody laughing about his little stupid remarks like that. Stuff like that went on you know.

And that ole Bill Corcoran that was so smart. Me and him kind of almost despised each other for a while. But then Bill would sit and tell what he was going to do when he got back to 'New Yok'. He talked… he had that New York drawl. And ser-

geant Fardie would set and tell, you know, what he was going to do when he got back to Boston. And so everybody watched after everybody. You grew up like that (snapping his fingers). You would go in as a kid and overnight you would become a man. If you didn't, you didn't last very long.

Glossary

ARP (Air Raid Precautions)   *Measures taken against air attack on British cities. Also the name of the government department in charge of air defense*

Allies   *Term generally used to describe the USA, British Empire and their allies in WW2 Artillery Heavy guns*

AAF (Army Air Forces)   *United States Army Air Forces. Its main role was to bomb German and Japanese cities in WW2*

Belsen   *Concentration camp in Germany that was liberated by the British in 1945*

Casualties   *People killed and/or wounded in war*

Concentration Camp   *Camps in Germany used by the Nazis to hold and torture their opponents, not the same as death camps. However, many died in such camps*

Convoy   *A ship, fleet, or group of vehicles, usually accompanied by a protecting escort*

D-Day   *Allied invasion of German occupied France, June 6, 1944*

Death Camp   *Camps for killing racial groups, especially Jews, which the Nazis considered to be inferior*

Luftwaffe   *German Air Force*

Mulberry Harbour   *An artificial floating harbor built to supply the D-Day landing forces in 1944*

Nazi Party   *Ruling political party in Germany 1933-45, headed by Adolf Hitler. (The National Socialist German Worker's Party)*

Omaha   *Codename for one of the beaches in the D-Day landings of 1944*

Panzer   *German tank*

RAF (Royal Air Force) *British Air Force*

RN (Royal Navy) *British Navy*

Sword *Codename for one of the beaches in the D-Day landings of 1944*

Torpedo *Missile fired by submarines, ships and aircraft, designed to sink other surface vessels or submarines*

U-Boat (Unterseeboot) *German submarine*

Utah *Codename for one of the beaches in the D-Day landings of 1944*

Notes

1. A Metronome at Sea:

[1] "A Trip Across Time," The Queen Mary, accessed July 30, 2020, https://www.queenmary.com/history/

[2] Eric Niderost, "Voyages to Victory: RMS Queen Mary's War Service," Warfare History Network, accessed on July 30, 2020, https://warfarehistorynetwork.com/2017/01/16/voyages-to-victory-rms-queen-marys-war-service/

2. St. Ives by the Sea:

[1] Evan Andrews, "D-Day's Deadly Dress Rehearsal: The fiasco was initially covered up to ensure the D-Day mission remained secret." The History Channel, Updated June 5, 2019 https://www.history.com/news/d-days-deadly-dress-rehearsal.

[2] "175th Infantry (5th Maryland)," 29th Division Association, July 30, 2020, https://29th-divisionassociation.com/29th-division-175th-infantry/

[3] John C. McAdams, *The Americans at Normandy*, New York, New York: Forge Books, 2004, 36-37.

[4] Joseph Balkoki, *Beyond the Beachhead*, Mecahnicsburg, Pennsylvania: Stackpole Books, 2005, 169.

3. Disappearance at Omaha:

[1] "The History of LSTs," LST Scuttlebutt, The Official of the United States LST Association, accessed July 31, 2020, https://www.uslst.org/history.

[2] Thomas Paone, "Protecting the Beaches with Balloons: D-Day and the 320th Barrage Balloon Battalion," Smithsonian Air and Space Museum, June 4, 2019, https://airand-space.si.edu/stories/editorial/protecting-beaches-balloons-d-day-and-320th-barrage-balloon-battalion.

[3] John C. McAdams, *The Americans at Normandy*, New York, New York: Forge Books, 2004, 36-37.

[4] Joseph Balkoki, *Beyond the Beachhead*, Mecahnicsburg, Pennsylvania: Stackpole Books, 2005, 169.

4. The Foxhole that Birch Built:

[1] Logan Nye, "This is how Hedgerows made the invasion of Normandy a living hell," March 9, 2017, https://www.wearethemighty.com/articles/this-is-how-hedgerows-made-the-invasion-of-normandy-a-living-hell.

[2] Joseph Balkoki, *Beyond the Beachhead*, Mecahnicsburg, Pennsylvania: Stackpole Books, 2005, 214.

[3] Ibid., 217.

5. Tar Heel Down:
[1] Joseph Balkoki, *Beyond the Beachhead*, Mecahnicsburg, Pennsylvania: Stackpole Books, 2005, 228.

[2] Ibid., 285-287.

[3] Joseph Conner, "A Grave Task: The Wartime Job Nobody Wanted," HistoryNet, accessed 8/11/2020, https://www.historynet.com/grave-task-men-buried-wartime-dead.htm.

6. A Star and a Tunnel:
[1] "What is a Bronze Star," Medals of America – Military Blog, December 15, 2019, https://www.medalsofamerica.com/blog/what-is-a-bronze-star/#:~:text=The%20Bronze%20Star%20Medal%2C%20or,honor%20while%20serving%20their%20country

[2] Ibid.

9. The Barn:
[1] "Immigration, German," Library of Congress, accessed 8/16/2020, https://www.loc.gov/teachers/classroommaterials/presentationsandactivities/presentations/immigration/german8.html.

[2] Erin Blakemore, "The Nazi breeding and infanticide program you probably never knew about," Timeline, December 18, 2017, https://timeline.com/the-nazi-breeding-and-infanticide-program-you-probably-never-knew-about-cc5cc7b82fdc

[3] "Gardelegen," United States Holocaust Memorial Museum, Holocaust Encyclopedia, https://encyclopedia.ushmm.org/content/en/article/gardelegen, accessed August 20, 2020.

The Author

Dr. John F. Birch is from southwest Ohio, where he has lived for most of his life. After completing graduate school at Miami University in Oxford, Ohio he married his wife, Debbie. They enjoy spending time with their daughter Kara, son-in-law Dustin, and granddaughter Kierstin. While a Ph.D. student at the University of Dayton (U.D.) he co-authored a theology text entitled "Religious Diversity and the American Experience" with his advising professor and classmates. John is an ordained minister and currently teaches at Indiana Wesleyan University (IWU), where he serves as Associate Professor and Division Chair of Liberal Arts.

www.ingramcontent.com/pod-product-compliance
Lightning Source LLC
Chambersburg PA
CBHW071328120626
46546CB00002B/483